# Advances in African Economic, Social and Political Development

**Series Editors**

Diery Seck, CREPOL—Center for Research on Political Economy, Dakar Yoff, Senegal

Juliet U. Elu, Morehouse College, Atlanta, GA, USA

Yaw Nyarko, New York University, New York, NY, USA

Africa is emerging as a rapidly growing region, still facing major challenges, but with a potential for significant progress – a transformation that necessitates vigorous efforts in research and policy thinking. This book series focuses on three intricately related key aspects of modern-day Africa: economic, social and political development. Making use of recent theoretical and empirical advances, the series aims to provide fresh answers to Africa's development challenges. All the socio-political dimensions of today's Africa are incorporated as they unfold and new policy options are presented. The series aims to provide a broad and interactive forum of science at work for policymaking and to bring together African and international researchers and experts. The series welcomes monographs and contributed volumes for an academic and professional audience, as well as tightly edited conference proceedings. Relevant topics include, but are not limited to, economic policy and trade, regional integration, labor market policies, demographic development, social issues, political economy and political systems, and environmental and energy issues.

All titles in the series are peer-reviewed. The book series is indexed in SCOPUS.

Hassan Qudrat-Ullah

# Exploring the Dynamics of Renewable Energy and Sustainable Development in Africa

A Cross-Country and Interdisciplinary Approach

 Springer

Hassan Qudrat-Ullah
York University
Toronto, ON, Canada

ISSN 2198-7262                  ISSN 2198-7270    (electronic)
Advances in African Economic, Social and Political Development
ISBN 978-3-031-48527-5          ISBN 978-3-031-48528-2    (eBook)
https://doi.org/10.1007/978-3-031-48528-2

This Springer imprint is published by the registered company Springer Nature Switzerland AG
The registered company address is: Gewerbestrasse 11, 6330 Cham, Switzerland

Paper in this product is recyclable.

*This book is dedicated to*
*Muhammad* صلى الله عليه وسلم
*(Peace be upon him)*
*(570 AD–632 AD)*
*Who single-handedly energized and brought*
*everlasting light to overcome the darkness of*
*this world.*

# Preface and Acknowledgements

Africa is a continent of immense potential and challenges, especially when it comes to electrification. According to the International Energy Agency, about 600 million people in Africa still lack access to electricity, and many more suffer from unreliable and expensive power supply. This situation hampers economic development, social welfare, and environmental sustainability in the region.

However, Africa also has abundant renewable energy resources, such as geothermal, solar, wind, hydro, and biomass that can provide clean, affordable, and reliable electricity to millions of people. Renewable energy can also reduce greenhouse gas emissions, enhance energy security, diversify the energy mix, create jobs, and foster innovation. Therefore, renewable energy is widely recognized as a key driver of sustainable development in Africa.

But how can renewable energy be effectively developed and utilized in Africa? What are the impacts of renewable energy on sustainable development in different contexts and settings? What are the challenges and opportunities of renewable energy transition in Africa? And what are the policy implications and recommendations that can facilitate and accelerate the deployment of renewable energy sources in Africa?

These are some of the questions that this book aims to answer. This book is the result of a collaborative research project that involved researchers from different disciplines and countries. The book provides an overview of the electrification challenges and opportunities in Africa, with a focus on renewable energy sources. The book also presents case studies of four African countries, namely Cameroon, Nigeria, Uganda, South Africa, and Tanzania that have different energy profiles, potentials, and contexts. The book analyzes the electricity demand and supply, the policy and technical solutions, the socioeconomic benefits and environmental impacts, and the barriers and gaps in renewable energy development and utilization in these countries. The book also suggests some policy implications and recommendations that can facilitate and accelerate the deployment of renewable energy sources in Africa.

The book is intended for a wide audience of researchers, policymakers, practitioners, students, and anyone interested in renewable energy and sustainable development in Africa. The book also provides references and links to further sources of information for those who want to delve deeper into the topic.

The book is divided into eight chapters. Chapter 1 introduces the concept of sustainable development and its relation to renewable energy. Chapters 2–6 present the case studies of Cameroon, Nigeria, Uganda, South Africa, and Algeria, respectively. Chapter 7 identifies some research gaps that remain to be addressed by future studies on renewable energy and sustainable development in Africa.

I hope that this book will provide a useful and informative overview of renewable energy's impact on sustainable development in Africa. I also hope that this book will stimulate further research on this topic and inspire more action toward achieving universal access to clean and affordable electricity in Africa.

I would like to thank all the contributors, reviewers, editors, and readers of this book for their valuable input, feedback, support, and interest. We also acknowledge the work and knowledge of the members of our review panels, many of which had to be done at short notice. Any errors or omissions are my responsibility.

Thanks to all the people at Springer, USA, especially Niko Hisako with whom we corresponded for their advice and facilitation in the production of this book.

Manju Ramanathan, Straive, prepared a camera-ready copy of the manuscript with her usual professionalism and cooperation, and we wish to record our thanks to her.

Finally, I am grateful to my family, Tahira (for her unwavering, incredible, and selfless support all the time), Anam (for her spiritual perspective on things around us), Ali (for sparing time from his kingdom), Umer (for his occasional smiles on my work and his work too), Umael (for bearing with me on being away and asking about his recitation of Quran practice), and my mother, Fazeelat Begum (for their sacrifice, support, and prayers all along despite all the odds)—all the very source of my inspiration and desire to embark on this journey. It would be unfair not to acknowledge the constant and consistent prayers and cares she extends to me, Saira Bano, my mother-in-law.

Toronto, Canada                                                            Hassan Qudrat-Ullah
November 2023

# Contents

# Chapter 1
# Introduction—Exploring the Dynamics of Renewable Energy and Sustainable Development in Africa

**Abstract** Renewable energy (RE) is a key solution to address the energy challenges and aspirations of the African people, who still suffer from low access to electricity and high dependence on traditional biomass. However, RE in Africa faces many technical, economic, social, environmental, institutional, and policy barriers that need to be overcome. This book provides a comprehensive and multidisciplinary perspective on RE in Africa, covering various aspects such as drivers, impacts, and solutions for RE in Africa. The book also presents various case studies from different African countries and regions, using various methods and tools for data collection and analysis. The book aims to fill the gaps in the existing literature and practice on RE in Africa and to contribute to the advancement of knowledge and practice on RE in Africa.

*What is in the readers of this chapter?* This chapter introduces the main theme and purpose of this book, which is to explore the dynamics of renewable energy and sustainable development in Africa. It explains the background and rationale for choosing this topic, as well as the objectives and scope of the book. It also outlines the expected outcomes and contributions of the book. By reading this chapter, you will gain a comprehensive and insightful overview of the book and its relevance to your interests and needs.

## 1.1  Background and Rationale

Africa is a continent with abundant natural resources, rich cultural diversity, and immense potential for development. However, Africa also faces multiple challenges, such as poverty, inequality, conflict, climate change, and energy insecurity that hinder its progress and prosperity. Access to electricity is a key driver of economic and social development, as well as a prerequisite for achieving the Sustainable Development Goals (SDGs). However, more than 600 million people in Africa still lack access to electricity, especially in rural areas, where the electrification rate is less than

© The Author(s), under exclusive license to Springer Nature Switzerland AG 2024
H. Qudrat-Ullah, *Exploring the Dynamics of Renewable Energy and Sustainable Development in Africa*, Advances in African Economic, Social and Political Development, https://doi.org/10.1007/978-3-031-48528-2_1

10% (World Bank, 2019). About 900 million people rely on traditional biomass for cooking and heating. Rural electrification (RE) is a major challenge for African countries, as it requires significant investments, technical expertise, institutional capacity, and policy support. These energy challenges have negative impacts on the health, education, income, and environment of the African people.

Renewable energy sources, such as solar, wind, biomass, and geothermal, offer a promising solution to address the energy needs and aspirations of the African people. Renewable energy sources are clean, abundant, and sustainable, and can contribute to reducing greenhouse gas emissions, enhancing energy security and independence, creating jobs and income opportunities, improving social welfare and quality of life, and fostering innovation and entrepreneurship. Renewable energy sources can also support the achievement of the African Union Agenda of Vision 2063 and the Sustainable Development Goals (SDGs), which provide a common framework and vision for the continent.

Therefore, RE is a strategic priority for African countries, as well as for regional and international organizations, such as the African Union (AU), the African Development Bank (AfDB), the International Renewable Energy Agency (IRENA), and the World Bank. These organizations have launched several initiatives and programs to support RE in Africa, such as the New Partnership for Africa's Development (NEPAD), the Program for Infrastructure Development in Africa (PIDA), the African Renewable Energy Initiative (AREI), the Scaling Up Renewable Energy Program in Low-Income Countries (SREP), and the Lighting Africa Program. However, RE in Africa still faces many gaps and challenges that need to be addressed (Foster and Briceño-Garmendia 2010; IRENA 2020).

There is a need for more evidence-based research and analysis on the drivers, barriers, impacts, and solutions for RE in Africa. There is also a need for more innovative and context-specific approaches and models for RE planning, implementation, and evaluation. Furthermore, there is a need for more stakeholder engagement and coordination among the public sector, the private sector, civil society, and the local communities.

This book aims to fill these gaps and contribute to the advancement of knowledge and practice on RE in Africa. The book provides a comprehensive and multidisciplinary perspective on RE in Africa, covering various aspects such as technical, economic, social, environmental, institutional, and policy issues. The book also presents various case studies from different African countries and regions, such as Cameroon, Nigeria, Uganda, South Africa, and Algeria. The book uses various methods and tools for data collection and analysis, such as literature reviews, surveys, interviews, focus groups, system dynamics modeling, scenario analysis, feasibility assessment, impact evaluation, and policy recommendations.

## 1.2 Objectives and Scope

The main objective of this book is to provide a comprehensive overview of the renewable energy sector in Africa, with a focus on Cameroon, Nigeria, Uganda, South Africa, and Algeria. The book aims to explore the challenges and opportunities of various renewable energy technologies in addressing the energy needs and aspirations of the African people. The book also aims to examine the socio-economic and environmental impacts of renewable energy projects, as well as their alignment with the African Union Agenda of Vision 2063 and the Sustainable Development Goals. The book will use a combination of theoretical and empirical methods to provide a comprehensive and rigorous assessment of the renewable energy sector in Africa.

The scope of this book covers five African countries: Cameroon (Central Africa), Nigeria (West Africa), Uganda (East Africa), South Africa (Southern Africa), and Algeria (North Africa). These countries were selected based on their geographical diversity, their different levels of electrification, their different sources of energy, and their different experiences with RE projects. The book focuses on four main sources of renewable energy for RE: solar, wind, biomass, and geothermal. These sources were selected based on their availability, potential, and suitability for rural areas in Africa.

## 1.3 Expected Outcomes and Contributions

The expected outcomes of this book are:

- To provide a comprehensive and multidisciplinary overview of RE in Africa.
- To provide evidence-based insights and recommendations for policymakers, practitioners, researchers, and stakeholders involved in RE in Africa.
- To showcase best practices and lessons learned from successful RE projects in Africa.
- To stimulate further research and innovation on RE in Africa.

The contributions of this book are:

- To fill the knowledge gap on RE in Africa by providing a comprehensive and multidisciplinary analysis of RE drivers, barriers, impacts, and solutions.
- To propose a paradigm shift in RE investments in Africa by focusing on demand-side factors, productive uses of electricity, and income generation.
- To apply various methods and tools for data collection and analysis, such as system dynamics modeling, scenario analysis, feasibility assessment, impact evaluation, and policy recommendations.
- To present case studies from different African countries and regions on RE planning, implementation, and evaluation.

## 1.4   Organization of the Book

The book is organized into seven chapters as follows:

- Chapter 1: Introduction: An Overview.
  This chapter provides the background and rationale, the objectives and scope, the expected outcomes and contributions, and the organization of the book.
- Chapter 2: Rural Electrification in Cameroon: Challenges, Opportunities, and a Proposed Model. This chapter analyzes the current status and challenges of RE in Cameroon and proposes a new model for RE planning and implementation based on a participatory and decentralized approach.

  *What is in for the readers of this chapter?* In Chap. 2, researchers can avail the identified future research on the factors influencing rural electrification in Cameroon and the effectiveness of the proposed model. Policymakers can use the stakeholder theory to design and implement rural electrification policies that are participatory and inclusive. Practitioners can learn from the challenges and opportunities of rural electrification in Cameroon and apply the best practices and lessons learned. University's research libraries and research centers can benefit from the comprehensive and updated literature review on rural electrification in Cameroon and Africa.
- Chapter 3: Energy Transition to Cleaner Energy in Nigeria: A Scenario-Based Modeling Approach. This chapter examines the energy transition and energy demand in the Nigerian household sector and uses a system dynamics model to explore different scenarios and policy options for enhancing RE development and utilization in Nigeria.

  *What is in for the readers of this chapter?* In Chap. 3, researchers can avail the scenario-based modeling method to explore the dynamics of the energy transition to cleaner energy in Nigeria and other countries. Policymakers can use the results and recommendations of the scenario analysis to formulate and evaluate policies and strategies for promoting liquefied petroleum gas as a substitute for traditional fuelwood in the household sector. Practitioners can learn from the bottlenecks and solutions for mainstreaming liquefied petroleum gas in the household sector and apply them in their contexts. University's research libraries and research centers can benefit from the data and information on the energy demand and supply in the Nigerian household sector and the potential of liquefied petroleum gas to reduce greenhouse gas emissions and improve health outcomes.
- Chapter 4: Feasibility of Bamboo Biomass Gasification for Rural Development in Uganda. This chapter assesses the technical, economic, social, and environmental feasibility of using bamboo biomass gasification for RE in Uganda, and provides recommendations for policy and practice.

  *What is in for the readers of this chapter?* In Chap. 4, readers can find a conceptual and methodological framework for studying bamboo gasification as a social innovation for sustainable energy and rural development in Uganda and other countries. Researchers can use the conceptual model and the 5-step approach to guide their research questions, hypotheses, data collection, and analysis methods.

Policymakers can use the findings and recommendations of the feasibility assessment to support and facilitate the promotion and commercialization of bamboo gasification projects. Practitioners can learn from the literature and case studies on bamboo cultivation and utilization, gasification technologies, syngas applications and markets, socio-economic and policy aspects, and sensitivity analysis. University's research libraries and research centers can benefit from the knowledge and insights on bamboo gasification and its potential and challenges.

- Chapter 5: Renewable Energy and Sustainable Development in South Africa: Challenges, Barriers, and Solutions. This chapter reviews the current status and trends of RE development and consumption in South Africa and identifies the drivers and barriers for RE adoption and diffusion. The chapter also assesses the impacts of RE on various dimensions of sustainable development and proposes policy and technical solutions for enhancing RE development and utilization in South Africa.

    *What is in for the readers of this chapter?* In Chap. 5, researchers can avail a comprehensive and updated review of the literature on renewable energy and sustainable development in South Africa, as well as identify some research gaps and limitations, and directions for further research. Policymakers can gain valuable insights and lessons on the policy and technical solutions for enhancing the role of renewable energy in advancing sustainable development in South Africa and beyond. Practitioners can learn about the current status and trends of renewable energy development and consumption in South Africa, as well as the drivers and barriers to renewable energy adoption and diffusion. University's research libraries and research centers can access useful references or resources on renewable energy and sustainable development issues. Students both undergraduate and graduate can enhance their knowledge and understanding of the impacts of renewable energy on various dimensions of sustainable development in South Africa, as well as the trade-offs and synergies between renewable energy and other development goals and priorities.

- Chapter 6: Renewable Energy Dynamics in North Africa: A Systems Thinking Approach Using Algerian Case. This chapter is about the dynamics of renewable energy (RE) in North Africa, with a focus on the case of Algeria. This chapter conducts a literature review of articles and journals on this topic and then designs a conceptual model to analyze the dynamics of RE in Algeria using systems thinking and causal loop diagrams. The chapter concludes with some policy implications and recommendations for enhancing RE development and integration in Algeria.

    *What is in for the readers of this chapter?* Chapter 6 offers the readers a comprehensive and insightful analysis of the renewable energy (RE) dynamics in North Africa, with a focus on the case of Algeria. The readers can learn about the energy resources and challenges in the region, such as the electricity demand and supply gap, and the potential and policies for RE development and integration. The readers can also understand how various factors, such as policy, technology, economics, behavior, and sustainability, influence RE development and integration in Algeria, using systems thinking and causal loop diagrams. Moreover, the readers can identify some of the feedback loops and interactions between these factors, which can create complex and dynamic behaviors in the RE system. Finally, the readers can explore some of the policy implications and recommendations for enhancing RE development and integration in Algeria, as well as overcoming some of the challenges and constraints that hinder it.

- Chapter 7: Conclusion. This chapter summarizes the main findings and contributions of the book and provides some directions for future research.

## References

Foster V, Briceño-Garmendia C (eds) (2010) Africa's infrastructure: a time for transformation. World Bank Publications

IRENA (2020) Electricity access in Sub-Saharan Africa: uptake, reliability, and complementary factors for economic impact. International Renewable Energy Agency, Abu Dhabi

World Bank (2019) Rural electrification: how much does Sub-Saharan Africa need the grid? Retrieved from https://blogs.worldbank.org/developmenttalk/rural-electrification-how-much-does-sub-saharan-africa-need-grid

# Chapter 2
# Improving Rural Electrification Access in Cameroon: A Qualitative Study

**Abstract** Cameroon's Vision 2035 aims to become an emergent economy, and energy is a crucial resource for achieving this goal. However, only 20% of the rural population has access to electricity, which limits their economic opportunities. This study investigates the reasons for the slow progress of rural electrification in Cameroon since the establishment of the Rural Electrification Agency in 1998. The study also aims to identify the challenges and propose a model for rural electrification. The study uses the stakeholder theory as a framework and adopts a survey research design with a qualitative approach. Data is collected from primary and secondary sources, using semi-structured interviews and personnel as data collection instruments. The study samples 10 key institutions and selects 7 respondents for data analysis, which is done through in-depth content analysis. The study finds that corruption and poor coordination of the rural electrification sector are the main barriers to rural electrification in Cameroon. The study recommends a bottom-up policy-making process for rural electrification and suggests a model for implementation.

*What is in for the readers of this chapter?* In this chapter, researchers can avail the identified future research on the factors influencing rural electrification in Cameroon and the effectiveness of the proposed model. Policymakers can use the stakeholder theory to design and implement rural electrification policies that are participatory and inclusive. Practitioners can learn from the challenges and opportunities of rural electrification in Cameroon and apply the best practices and lessons learned. University's research libraries and research centers can benefit from the comprehensive and updated literature review on rural electrification in Cameroon and Africa.

## 2.1 Introduction

Cameroon has a wealth of energy resources, such as oil, natural gas, bauxite, forestry, hydropower, wind, solar, biomass, and geothermal. However, these resources are not fully exploited and developed, especially renewable ones. The main sources

© The Author(s), under exclusive license to Springer Nature Switzerland AG 2024
H. Qudrat-Ullah, *Exploring the Dynamics of Renewable Energy and Sustainable Development in Africa*, Advances in African Economic, Social and Political Development, https://doi.org/10.1007/978-3-031-48528-2_2

of commercial energy in Cameroon are hydropower for electricity generation and petroleum products for transportation. Biomass, mainly in the form of traditional biomass, dominates energy consumption in the country, particularly in rural areas where modern energy services are scarce (Buzanakova 2014b, a). The urban population enjoys relatively high access to electricity (between 65 and 88%), while the rural population suffers from very low access (about 14%). Cameroon has various renewable energy sources that could be used to improve its energy situation (IEA 2014a, b, c, d, 2017, 2019; IsDB 2017a). Hydropower is the main source of electricity generation in Cameroon, accounting for 74.9% of the electricity mix and having an installed capacity of 1558 MW in 2009. In 2015, the total electricity generation was 6758 GWh and the sectoral consumption was 5784 GWh. The industry sector was the largest electricity consumer with 55.2%. The government has set ambitious targets to increase the electrification rate in the country by 2020: 48% at the national level, 75% at the urban level, and 20% at the rural level (REEP 2013). The high dependence on biomass in Cameroon is due to the lack of modern energy sources and the inability to afford them.

Energy access is essential for achieving development goals in Africa, which has the lowest levels of energy access and human development in the world (UNDP 2012). Poor electricity services are considered a major obstacle to socio-economic development in Africa (Onyeji et al. 2012). Energy in Cameroon plays a vital role in shaping its economy and its vision to become an emerging economy by 2035. The country has abundant reserves of oil and natural gas and is pursuing new policies to improve and diversify its energy sources. The increasing global competition has also led to the expansion of the energy sector in Cameroon. Energy in Cameroon comprises its oil and natural gas reserves, hydroelectric energy, and other renewable resources.

Despite the moderate levels of electricity access in Cameroon compared to neighboring countries, very few rural facilities (schools, clinics, businesses, etc.) are electrified. Electrification is mainly achieved through grid extensions, which are often costly and time-consuming as they involve only government initiation, development, and implementation. This lengthy government process involves adopting laws and policies, allocating and approving budgets in parliament, awarding tenders and contracts which are often corrupt in the African context, delaying effective implementations for political reasons, and setting up special agencies to manage such processes which are still controlled from above. A typical example of such an agency is the Rural Electrification Agency (REA) in Cameroon which was created through Law No. 98/022 of December 24th, 1998 governing the electricity sector in Cameroon. The REA is responsible for promoting and developing rural electrification across the national territory in conjunction with the administrations, and public and private agencies concerned. It is supposed to contribute to the elaboration and implementation of government policy in the domain of rural electrification in Cameroon (Agence de l'Electrification Rurale du Cameroun (AER) 2015; REA 2015).

The government established the Rural Electrification Agency (REA) in 1998 to promote and implement rural electrification projects across the country, but the progress has been slow and uneven. This study aims to address the research question: Why is there still a low rate of rural electrification in Cameroon and how can it be improved? The objectives of this study are to:

- Analyze the current status and challenges of rural electrification in Cameroon,
- Review the best practices and lessons learned from other countries that have successfully implemented rural electrification projects,
- Propose a model for improving rural electrification in Cameroon that is based on stakeholder involvement, policy coherence, resource optimization, and
- Provide practical and policy recommendations for enhancing rural electrification in Cameroon.

This study is novel and innovative in several ways. First, it proposes a model for improving rural electrification in Cameroon that takes into account the specific characteristics and challenges of the country's energy sector, such as the diversity of energy resources, the low access to modern energy services, the corruption and inefficiency of the government processes, and the lack of involvement of the private sector and the local communities. Second, it applies a mixed-methods approach that combines quantitative and qualitative data collection and analysis to assess the current situation of rural electrification in Cameroon and to identify the best practices and solutions from other countries that have successfully implemented rural electrification projects. Third, it provides practical and policy recommendations for enhancing rural electrification in Cameroon that are based on evidence and stakeholder engagement. This study contributes to the literature on rural electrification in Africa (e.g., How We Made It In Africa 2017; World Bank 2019; FUSS, Cameroon, and Renewable Energy 2022; Cairn International 2016) and to the achievement of sustainable development goals related to energy access, poverty reduction, and environmental protection.

## 2.2   Background Information and Literature Review

### 2.2.1   Electricity Situation in Cameroon

Cameroon is a country on the rise, with a booming economy and a growing electricity demand. Over the last ten years, its economy has grown by an impressive 5.9% in 2015, making it one of the fastest-growing in Africa (Iweh et al. 2023). Along with this economic growth, more and more people are enjoying the benefits of electricity in their homes and businesses. The national access to electricity has risen from 37% in 1996 to 48% in 2007, surpassing the average for African countries with abundant natural resources (AfDB 2021). Electricity is especially accessible in urban areas, where between 65 and 88% of the population has access to it (Chaurey and Kandpal 2010; Iweh et al. 2023). However, rural areas still lag, with only about 14% of the population having access to electricity (AfDB 2021). Cameroon relies heavily on hydropower to generate electricity, which accounts for 70% of its total production (AfDB 2021). Hydropower is a clean and renewable source of energy, but it also depends on the availability and quality of water resources (Conyers 1985; Crousillat et al. 2010; Iweh et al. 2023). Cameroon faces many challenges and opportunities to improve its electricity sector and meet the needs of its growing population.

## 2.2.2  *Electricity Demand*

Cameroon is a country with a growing appetite for electricity, driven by its economic development and population growth. In 2015, the country consumed 5784 GWh of electricity, of which 20.1% was used by households, 23.2% by commercial service, and 55.2% by the industrial sector (Iweh et al. 2023). However, this consumption was not evenly distributed across the country, as shown in Fig. 2.1.

One of the main consumers of electricity in Cameroon is the aluminum smelter at Edéa ALUCAM, which accounts for nearly 60% of the total production, leaving just 40% for the rest of the population (Fotsing et al. 2014). This means that many regions and sectors face electricity shortages and blackouts, especially during peak periods. The lowest covered region in electricity in 2007 was the Far North, while the highest coverage rate was recorded in the South. The urban areas of Yaoundé and Douala had the most favorable coverage rate, with an average of 48.3% in the same year, which was 3.9 times higher than in rural areas (Liang et al. 2017; Fotsing et al. 2014). The poorest regions, with the most critical poverty index, appeared to be underserved in electrical energy. These regions were: the Great Northern regions (Far North and North) and East region.

To project the future electricity demand in Cameroon, various scenarios have been developed based on the trends of the past ten years and the expected economic and demographic growth. One of these scenarios is the business as usual (BAU) scenario, which assumes no major changes in the electricity sector policies and practices. Figure 2.2 shows the consumption projections under this scenario for different sectors (industry, tertiary buildings, residential, and households) from 2015 to 2035.

According to this scenario, the electricity demand in Cameroon is expected to increase from 5.41 TWh in 2015 to 19.79 TWh in 2035, with an average annual growth rate of 6.8% (Iweh et al. 2023). The industrial sector would remain the largest consumer of electricity, followed by the residential sector. The supplier of electricity (ENEO), responsible for the overall electricity system management (production, transmission, and distribution), is currently struggling to meet the growing power

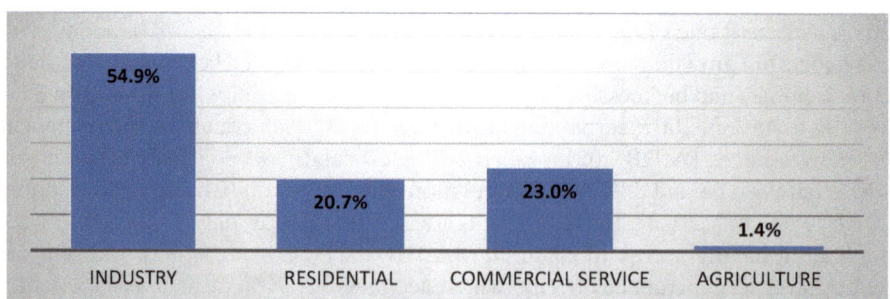

**Fig. 2.1** Cameroon electricity consumption by sector in 2014 (IEA 2014a)

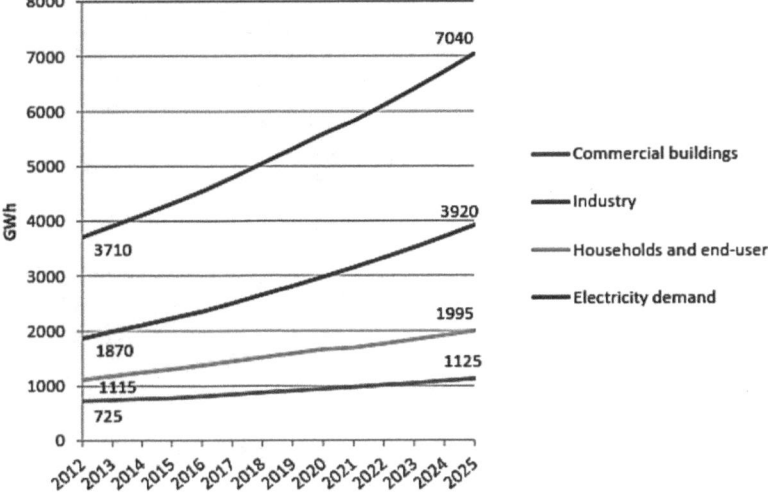

**Fig. 2.2** Consumption projections on BAU (business as usual) baselines (Fuss 2013)

demand but is incapable of satisfying all needs, especially during peak periods (Guefano et al. 2020). Therefore, there is a need for more investment and innovation in the electricity sector to ensure a reliable and sustainable supply of electricity for all.

## 2.2.3 Electricity Production

Electricity production: Cameroon is a country with abundant natural resources for electricity production, especially hydropower, but it still faces challenges in providing universal access and reliable service to its population. The country's electrification rate is moderate, at 55% in 2013, but it varies widely across regions and sectors, with 10 million people without access to electricity (Iweh et al. 2023). The urban and rural electrification rate was 88 and 17% in 2016, respectively (AfDB 2021). In 2015, the total electricity production was 6758 GWh, with a total installed capacity of around 1925.86 MW (on-grid and off-grid) (Fotsing et al. 2014). Figure 2.3 shows the breakdown of electricity production by source in 2015.

Hydropower dominates electricity generation in Cameroon, with 74.9% of the total production, followed by self-production (mainly thermal) with 22%, with an installed capacity of 1558 MW in 2009 (Fotsing et al. 2014). Apart from hydropower, which has a large potential for expansion, Cameroon also has other renewable energy sources, such as biomass, solar, and wind, which could contribute to diversifying its electricity mix and reducing its greenhouse gas emissions. According to Lighting Africa Policy Report (2012), the introduction and proper implementation of off-grid, low-cost, reliable, and durable lighting options (including renewables) in Cameroon could result in rapid growth in electrification rates and energy access in rural and

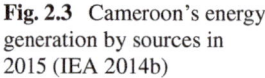

**Fig. 2.3** Cameroon's energy generation by sources in 2015 (IEA 2014b)

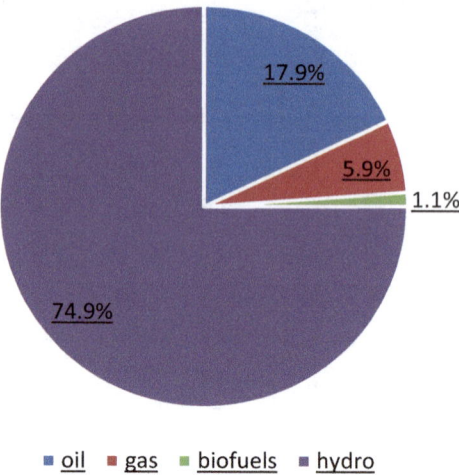

urban areas between 2010 and 2025. Figure 2.4 illustrates the model growth of electricity access in Cameroon under this scenario.

Cameroon has an on-grid total installed capacity of around 1323.96 MW, of which approximately two-thirds is hydropower and the rest is thermal (604.96 MW) (Iweh et al. 2023). Cameroon has three large-scale hydropower plants with a total capacity of 719 MW: Song Loulou (384 MW), Edéa (263 MW), and Lagdo (72 MW) combined with three upstream reservoirs on tributaries to the Sanaga River (Mbakaou, Mape, Bamendjin) with a total storage capacity of 7.6 billion cubic meters (Fotsing et al. 2014). Figure 2.5 shows the electricity generation by source in Cameroon in 2016.

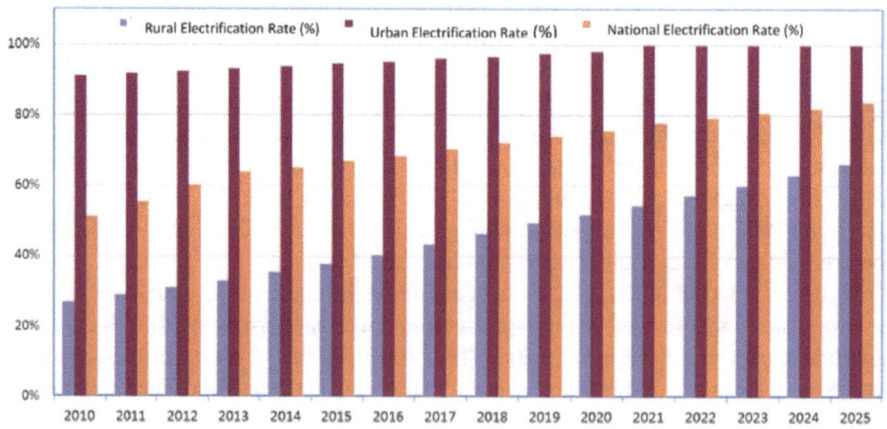

**Fig. 2.4** Model growth of electricity access in Cameroon between 2010 and 2025 (Lighting Africa 2012)

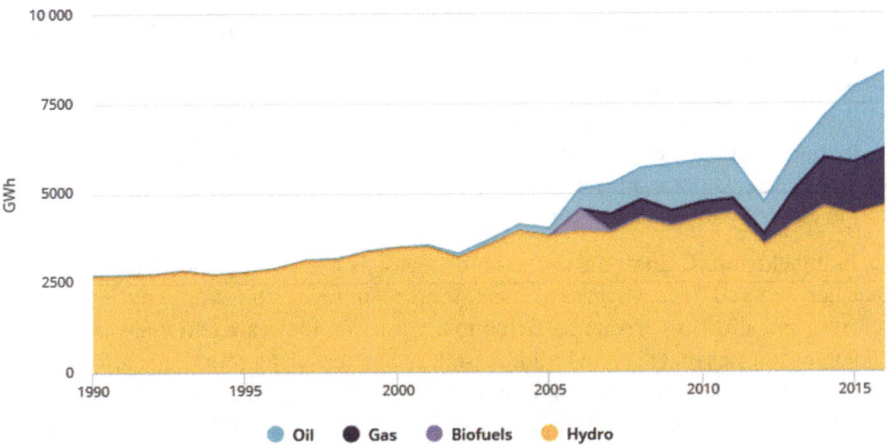

**Fig. 2.5** Electricity generation by source (IEA 2016)

In addition to the existing hydropower plants, Cameroon has several projects under construction or planned to increase its hydropower capacity, such as Lom Pangar (30 MW), Memve'ele (211 MW), Mekin (15 MW), Nachtigal (420 MW) and Song Dong (270 MW) (AfDB 2021). However, hydropower is also vulnerable to climate variability and water scarcity, which could affect its reliability and sustainability. Therefore, Cameroon needs to diversify its electricity sources and invest in other renewable energy technologies, such as biomass, solar, and wind. In 2010, an additional 586 MW of thermal capacity was installed for self-generation, of which 562 MW are onshore and 24 MW are offshore (Fotsing et al. 2014). Furthermore, there are 26 isolated diesel grids with a total installed capacity of 15.3 MW and a total power output in 2011 of 42,765 MWh. Several micro- and pico-hydropower projects with a total installed capacity of 515.5 kW have also been developed by Action pour un Développement Équitable, Intégré et Durable (ADEID). The country's installed capacity stands at 1442 MW (Business in Cameroon 2018).

### 2.2.4  Electricity Sector Structure in Cameroon

Cameroon's electricity sector has undergone several changes since 1998 when the government started to reform it. The Electricity Law of 1998 and the Electricity Decree of 2000 allowed the state-owned power company, SONEL, to be privatized and liberalized. In 2001, AES bought a majority stake in SONEL and formed AES SONEL. AES SONEL and the government signed a concession agreement to regulate their relationship (Castalia 2015). In 2014, AES sold its shares in AES SONEL to Actis, who renamed the company ENEO Cameroon. ENEO Cameroon runs three separate grids that serve different parts of the country:

- The Southern Interconnected Grid: This grid connects the main hydropower plants to the large aluminum factories and the biggest cities of Yaoundé and Douala, where most of the electricity is consumed,
- The Northern Interconnected Grid: This grid distributes the electricity produced by the Lagdo power plant to meet the low demand of the region,
- The Eastern Interconnected Grid: This grid is a low-voltage network that serves the eastern part of the country.

In conclusion, Cameroon's electricity sector is a complex and dynamic system that has evolved. The government has tried to improve the sector by introducing reforms and allowing private participation. However, there are still many challenges and opportunities that need to be addressed, such as ensuring reliability, affordability, and sustainability of electricity supply, expanding access to rural areas, and promoting renewable energy sources (Djoedjom and Zhao 2018; Fowler 1991; Iweh et al. 2023). Cameroon's electricity sector is a key factor in the country's economic and social development, and it requires continuous monitoring and evaluation to ensure its optimal performance (Nkongho et al. 2002).

### 2.2.5  Other Important Electricity Sector Actors

Besides ENEO, the main entities in the power sector in Cameroon include:

- Ministry of Water Resources and Energy (Ministère de l'Eau et de l'Energie, MINEE): The MINEE is responsible for implementing government action in the energy sector and monitoring energy sector activities (Renewable Energy and Energy Efficiency Partnership (REEP) 2013; Nkongho et al. 2002).
- Rural Electrification Agency (Agence d'Electrification Rurale–AER): AER promotes and develops rural electrification projects across the country by providing financing to communities and operators. AER is a public institution established by Decree in 1999 (Decree 99/193 of 8th September 1999). AER is under the technical authority of the MINEE and the financial supervision of the Minister of Finance. In 2013, a Decree of the President of the Republic reinforced the AER mission (Decree no. 2013/204 of 28th June 2013). AER's tasks and responsibilities were identified more precisely in the decree's section on rural electrification (Castalia 2015).
- Rural Energy Fund (Fond Énergie Rurale—REF): REF was created by presidential decree in 2009. It is not a legal entity and is managed by AER. The main purpose of REF is to grant a partial subsidy to priority investment programs involving rural electrification (McElroy and Mills 2000; Castalia 2015).
- Electric Sector Regulation Agency (Agence de Régulation du Secteur de l'Electricité–ARSEL): ARSEL was established by the 1998 Electricity Law governing the electricity sector. Decree No. 99/125 of 15th June 1999 organizes the functioning of ARSEL in regulating the electricity sector. ARSEL ensures fair competition, protects consumer rights, and monitors tariff adjustments (Nkongho et al. 2002).

- Electricity Development Corporation (EDC): EDC is a state-owned company created by decree (Decree no. 2006/406 of 29th November 2006) that develops the electricity sector including all hydroelectric projects in the country. EDC is mainly in charge of building and operating dams as well as the operation and maintenance of storage dams (barrage-réservoirs) (Castalia 2015).
- The Committee of Planning and Programming of Rural Energy (COPPER): The Committee of Planning and Programming of Rural Energy (Comité de Planification et Programmation de l'Energie Rurale) was established by presidential decree in 2009. It has as its main purpose is to ensure the proper allocation of resources and subsidies administered by the FER. The COPPER is chaired by the Minister of Water and Energy (Castalia 2015). The power market structure in Cameroon is presented in Fig. 2.6.

The electricity sector in Cameroon faces several challenges, such as inadequate generation capacity, high transmission losses, low electrification rate, poor quality of service, and environmental impacts. To address these challenges, the government has adopted several policies and strategies, such as the Electricity Sector Development Plan (PDSE), the National Energy Policy (PNE), the Rural Electrification Master Plan (PDER), and the Renewable Energy Development Strategy (SDER). These policies aim to increase electricity access, diversify energy sources, improve efficiency, and promote private-sector participation (Iweh et al. 2023).

Overall, in this section, we have discussed the main actors and institutions involved in the electricity sector in Cameroon, as well as their roles and responsibilities. We have also presented the power market structure and the challenges faced by the sector.

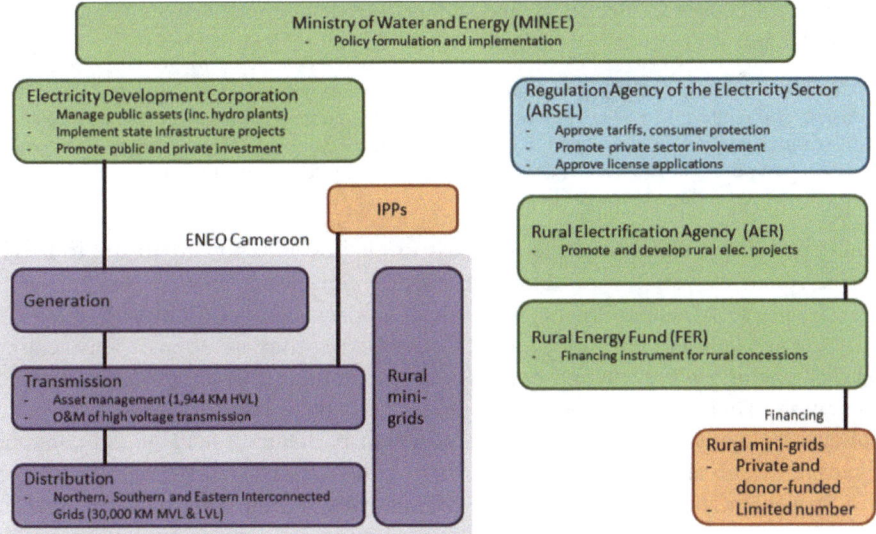

**Fig. 2.6** Cameroon power market structure (*Source* Adapted from Cascadia 2015)

We have highlighted some of the policies and strategies adopted by the government to improve the performance and sustainability of the sector. In the next section, we will analyze the current situation of rural electrification in Cameroon.

### 2.2.6  Rural Electrification in Cameroon

Rural electrification is the process of providing electricity access to households or villages in remote areas. It includes the provision of electricity to areas with highly dispersed potential consumers and low demand (Niez 2010). Rural electrification may include village-level input into the social and economic development, utilized by households, farms, and establishments. It is likely to lead to socio-economic changes such as employment, income, and productivity, especially in agriculture.

Research in rural electrification shows the potential for private and semi-private actors to enhance access to energy in poor remote, inaccessible areas. The poor returns and technical difficulties in rural electrification call for the need to establish an autonomous division within the utility and devolve more responsibility to the local organizations such as local communities and cooperatives (UN General Assembly 2015; UNIDO 2016; Hisham 2010). Here, the author highlights an important role the private and semi-private actors can play in rural electrification.

However, rural electrification also faces several challenges, such as the high cost of grid extension, low reliability and quality of supply, inadequate clean energy usage, poor EV charging infrastructure, and high upfront costs for electrification projects (World Bank Group 2018; Qmerit 2022). These challenges hinder the achievement of universal access to electricity and sustainable development goals in rural areas (Figs. 2.7 and 2.8).

> To overcome these challenges, various solutions have been proposed and implemented, such as off-grid renewable energy systems, mini-grids, stand-alone systems, hybrid systems, and innovative financing mechanisms (Deloitte 2023). These solutions offer advantages such as lower cost per connection, faster deployment, reduced environmental impact, increased local participation, and improved resilience.

Despite the moderate electricity access levels in Cameroon compared to neighboring countries, very few rural facilities (schools, clinics, businesses, etc.) are electrified and electrification is mainly through grid extensions (Suhlrie et al. 2018; Sutcliffe and Court 2006; Cascadia 2015). Grid extensions are often expensive, time time-consuming as they involve only government initiation, development, and implementation. This lengthy government process mostly involved adopting laws and policies, budgetary allocations and adoptions in parliament, tenders and contracts awarding which are most often very corrupt in the African settings, delays in effective implementations for political reasons, and the setting up of special agencies to manage such processes of which powers are never given to the agency to act directly still controlled from above. A true example of such an agency is the REA in Cameroon in charge of the development and implementation of rural electrification projects of diverse origins which receives instructions from the Ministry of Water and Energy Resources of Cameroon.

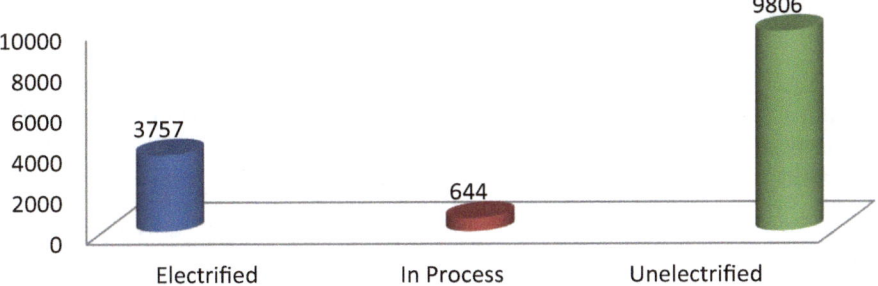

**Fig. 2.7** Rural electricity situation in the 14,207 identified Localities (*Source* Adapted from the Rural Electrification Master Plan 2016)

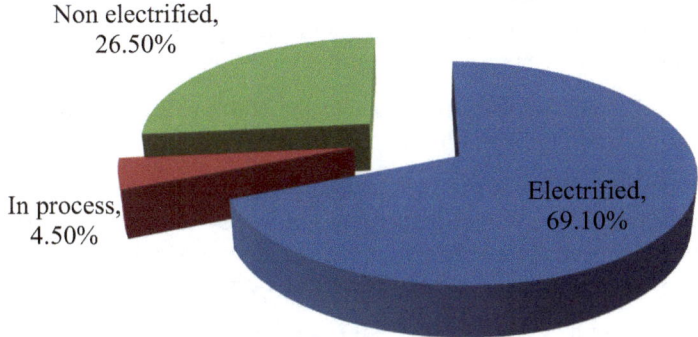

**Fig. 2.8** Electricity access rate for electrified localities (*Source* Adapted from the Rural Electrification Master Plan 2016)

Therefore, there is a need for more innovative and inclusive approaches to rural electrification that can leverage the potential of renewable energy sources, engage the private sector and local communities, and create enabling policy and regulatory frameworks. Such approaches can enhance the benefits of rural electrification for socio-economic development, poverty reduction, gender equality, health improvement, education enhancement, and environmental protection (United Nations Center for Human Settlements (UNCHS) 1984; United Nations Environment Programme (UNEP) 2001; ARE 2015).

In summary, we have discussed the main actors and institutions involved in the electricity sector in Cameroon, as well as their roles and responsibilities. We have also presented the power market structure and the challenges faced by the sector. We have highlighted some of the policies and strategies adopted by the government to improve the performance and sustainability of the sector. In addition, we have explored some of the solutions and opportunities for rural electrification that can address the gaps and barriers in the current system. We have argued that rural electrification is not only a technical issue but also a social and economic one that requires a holistic and participatory approach. In the next section, we will analyze the mission of the REA in Cameroon.

## 2.2.7   Mission of the REA

The Agency's mission is to promote and develop rural electrification in Cameroon, in collaboration with the relevant administrations, and public and private agencies. It contributes to the formulation and implementation of the Government's policy in the field of rural electrification. Specifically, it has the following objectives:

- Approve the plans and projects for rural electrification initiated by regional and local authorities;
- Popularize and promote renewable energy sources;
- Propose measures to attract investors in the field of rural electrification in Cameroon;
- Collect and disseminate information on various investment opportunities in the field of rural electrification;
- Conduct surveys and studies to find economically feasible technical solutions for rural areas, following approved standards and norms (AER 2015).

To achieve its mission, the REA created the Rural Electrification Master Plan (PDER) two years after its establishment, with the following assets:

- A comprehensive diagnosis of the electricity market in Cameroon;
- Identification, analysis, and evaluation of optimal systems to supply rural needs;
- More than 100 projects studied with a high productivity rate, involving: Network extensions; solar energy; mini hydropower; fuel-efficient generators; and biomass.

The PDER contains studies that were financed by the African Development Bank (AfDB) and the World Bank-funded National Energy Action Plan for Poverty Reduction (PANERP). The project aims to strengthen and extend power transmission and distribution systems to 423 new localities with almost 335,000 new consumers, especially rural dwellers. The total project cost was estimated at US$ 58.99 million and its implementation spanned 60 months from 2010. This project was sponsored by the African Development Fund (ADF), the Japanese International Cooperation Agency (JICA), and the Cameroon government (Ministère de l'Energie et de l'Eau (MINEE) «Présentation de AER», Moser 1989; Law no. 1999; AfDB 2009a, b). The government of Cameroon also received funding from the Islamic Development Bank (IsDB) to cover the costs of the rural electrification project, Phase II (IsDB 2017b). The government intends to use part of the funds for payments for the construction of medium and low-tension power distribution lines and the connection of the beneficiary households. It aims at improving the electrification rate to 98% from the current 18%, at a total cost of 600 billion CFA Francs.

Rural electrification in Cameroon is a key component of the country's development strategy, as it can improve the living conditions of rural communities, reduce poverty, enhance economic activities, and mitigate climate change impacts. According to a recent study by Njomo et al. (2021), rural electrification can also contribute to achieving several Sustainable Development Goals (SDGs), such as SDG 1 (no poverty), SDG 5 (gender equality), SDG 7 (affordable and clean energy), SDG 8 (decent work and economic growth), SDG 10 (reduced inequalities), SDG 13 (climate action), and SDG 17 (partnerships for the goals). Therefore, it is essential to support and monitor the progress of rural electrification projects in Cameroon, as well as to address the challenges and barriers that may hinder their success (Weible and Carter 2017; Wilson 1995; Young and Quinn 2012).

### 2.2.8 Stakeholder Participation

Due to the plethora of shortcomings deriving from top-down development efforts, participation has come to be recognized as an absolute imperative for development (Javadi et al. 2013; Jepsen and Eskerod 2009; Lekunze 2001a, b). Nevertheless, the concept of participation has remained an elusive one. Brohman (1996) posits that it has been given multiple meanings and connected to multiple methods of implementation in the last few decades. To him, there still exist many unanswered questions about who participates, what they participate in how they participate, and for what reasons they participate. Participatory policymaking is a general approach with the overall goal being to facilitate the inclusion of individuals or groups in the design of policies via consultative or participatory means to achieve accountability, transparency, and active citizenship (Liang et al. 2017; Rietbergen-McCracken 2011a, b). This push for participation can either take a top-down or bottom-up approach. This participatory process can entail seven different levels (Too and Weaver 2014; Tsoukiàs et al. 2013; Karl 2002a, b);

- Contribution: voluntary or other forms of input to predetermined programs and projects.
- Information sharing: Stakeholders are informed about their rights, responsibilities, and options.
- Consultation: stakeholders are allowed to interact and provide feedback, and may express suggestions and concerns. However, analysis and decisions are usually made by outsiders, and stakeholders have no assurance that their input will be used.
- Cooperation and consensus building: Stakeholders negotiate positions and help determine priorities, but the process is directed by outsiders.
- Decision making: Stakeholders have a role in making decisions on policy, project design, and implementation.
- Partnership: Stakeholders work together as equals towards mutual goals.
- Empowerment: transfer of control over decision-making and resources to stakeholders.

The participatory process can equally be a once-off exercise for a particular policy process or part of a systemic participatory governance approach by the organization/government in question. Permanent structures such as committees that include citizens' groups, community members, etc. can equally be involved. The policy itself can be local, national, or international (Midgley 1986; Rietbergen-McCracken 2011a, b). Participation is a complex multidimensional concept involving different stakeholders. Fleming (1991a, b) suggests that participation emphasizes the decision-making role of the community. At the community level, participation helps to improve the design of policies so that they correspond to the needs and conditions of the people to whom they are directed (McElroy and Mills 2000; Cornia et al. 1987a, b). Fenster (1993a, b) distinguishes between economists' definition of community participation, which is the equitable sharing of the benefits of projects; and social planners' definition as a community's contribution to decision-making. A much more realistic interpretation of community participation is given by Paul (1986:2) which he sees to be "an active process by which beneficiaries influence the direction and execution of a development project to enhance their well-being in terms of income, personal growth, self-reliance or other values they cherish". This therefore brings us to the question of participation, as either induced, or spontaneous. However, caution should be given to the frequently abused term participation because the bottom-up approach in itself has several limitations. While many development programs have been promoted by rhetoric about decentralization and participation, in practice, they have generally been either tightly controlled by the state or outside development institutions. Most states still fear that grassroots organizations (especially the youth) will generate popular empower.

## 2.3  Methodology

This study adopted a qualitative and exploratory research design, as it aimed to conduct in-depth studies and analyses of the issues related to stakeholder participation in policy-making for rural electrification in Cameroon. The study used purposive sampling to select key respondents who had concise and appropriate knowledge relevant to the topic. The data collection methods included interviews and personal observations as primary sources, and reviews from books, journal articles, blogs, and websites as secondary sources. The study population consisted of key government ministries and stakeholders involved in the energy sector, such as the REA, Ministry of Water Resources and Energy, National Financial Credit, Nkong Credit for Development (NC4D), Schneider Electric, Yandalux Cameroon Sarl, Globeleq Cameroon, GIZ Cameroon, Sinohydro Corporation Ltd., and SNV Cameroon. A sample size of 10 respondents representing these institutions was chosen for this study, as they were able to provide credible and factual information on the rural electrification sector. The data were analyzed through in-depth content analysis.

### 2.3.1  Theoretical Framework

To better understand the impediments and dynamics of successful rural electrification policy decision making, our theoretical framework is based on the stakeholder theory and evidence-based decision making as seen in Fig. 2.9.

The theoretical model assumes that blending stakeholder participation with evidence-based decision-making will lead to effective policymaking and further lead

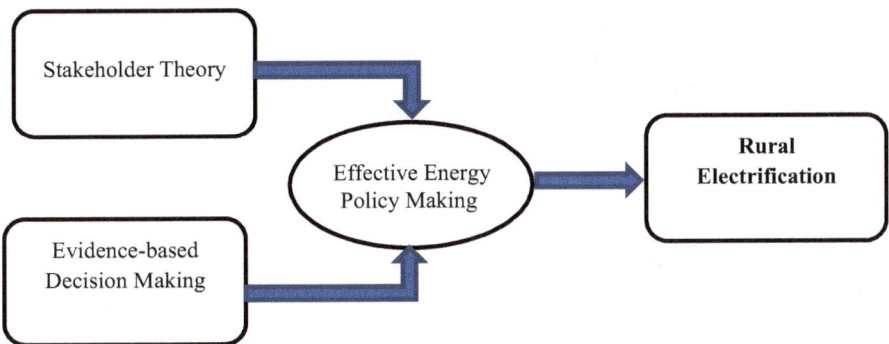

**Fig. 2.9** Theoretical framework of the study

to an improved rural electrification status. Stakeholder theory suggests that policy-makers should consider the interests and expectations of all relevant actors who are affected by or can influence the policy outcomes. Evidence-based decision-making implies that policymakers should use the best available scientific data and information to inform their choices and actions (Njomo et al. 2021; Rehman and Deyuan 2018). By combining these two approaches, we aim to identify the key factors and challenges that affect rural electrification policy design and implementation in Africa, as well as the potential solutions and best practices that can enhance the policy effectiveness and impact. We also seek to understand how different stakeholder groups perceive and evaluate the rural electrification policies and programs in their respective contexts, and how their feedback can be incorporated into the policy cycle.

## 2.4  Results and Discussion

### 2.4.1  Limited Co-ordination of the Rural Electrification Sector in Cameroon

The rural electricity sector in Cameroon involves many actors with different roles and interests. One would expect that such a diversity of actors would facilitate the rural electrification process, but this is not the case, as one respondent explains:

- *There is a lack of coordination. If coordination was good, we would do better. But since we have laws that we do not respect, coordination has become difficult. That is why we have poor jobs being done, and that is why we have double budgeting for certain localities. The law mandates the Rural Electrification Agency to oversee the rural electrification activities of other partner ministries. Some ministerial departments, however, do not respect this law for they see our service as an inferior establishment (Interview with R5, 7th May 2022).*

- The Rural Electrification Agency faces the challenge of coordinating the sector and this has been a major obstacle to the rural electrification progress of the country. The lack of enforcement mechanisms to compel all relevant sectors to adhere to regulations is making the task of the REA difficult.

## 2.4.2   Lack of Powers by the REA to Make Rural Electrification Policies

The creation of the REA in 1998 was seen as a potential solution to the problem of darkness in rural Cameroon. However, the REA only has the responsibility to implement rural electrification policies, not to make them. The challenge arises when the REA has to implement policies made by another institution, without having a say in the strategies used to design these policies. When asked about the parameters considered in the drafting of the Rural Electrification Master Plan, a respondent said;

> The priority for a place to be electrified in Cameroon is if it is an administrative Unit. So all divisional headquarters must be electrified after the Regional headquarters and then sub-divisional headquarters. There are some sub-divisional headquarters that are too far from the grid and those are not electrified. When you compare the cost of electrifying those sub-divisional headquarters, it becomes too expensive………The next policy is the potential for development in an area. (Interview with R5, 7th May 2022)

From these excerpts, it is clear that the main motivation for electrification is political, rather than economic. This explains why some administrative units with low economic potentials are electrified in Cameroon, while other localities with better economic prospects are left out. This aspect of political motivations is further confirmed by another respondent who said;

> The Rural Electrification Agency doesn't make rural electrification policies. The sector has stakeholders and policy is developed or elaborated by the Ministry of Water and Energy…….The Presidency is involved, the Prime Ministry is involved and other sectorial ministries are involved. The REA is now charged with the implementation while ARSEL is the regulatory agency. (Interview with R5, 10th May 2022).

In such a situation, policymaking is done by another institution, and implementation is done by another institution. This makes the policy process complex and cumbersome. The policy process could be more effective if policymakers were also involved in implementation, or vice versa. With the REA only charged with the implementation of existing policies, there is a limited sense of policy ownership and commitment to ensure policy success. Besides, the fact that policymaking is influenced by the presidency and other ministerial departments indicates that the policy-making process is politicized. This suggests that policymaking is not based on evidence, but on other considerations, and that the expected results of policies are not backed by research findings and facts.

### 2.4.3  Lack of Funds

Implementing electrification projects requires a lot of financial resources, and most Sub-Saharan African countries like Cameroon often depend on aid and loans to carry out these projects. For Cameroon which relies more on building dams to exploit hydropower, these projects are very costly to establish. A staff of the Rural Energy Fund explained the issue of funding in detail:

> The first challenge is funding. When you look at all the projects the Rural Electrification Agency is supposed to implement, most of them rely on foreign donors to succeed. The Agency itself relies on state funds to pay its staff and cover other recurrent expenditures. No matter how fast we want to go, we can only implement projects for which the government, donors, and lenders provide funding…..Cameroonian investors and businessmen do not love investing long term, what they prefer is projects with quick turnover. (Interview with R6, 7th May 2022).

Other ministries are supposed to submit their projects to the REA to avoid double funding of localities and this can also help to prevent financial waste. This is a regulation that is however not respected. While acknowledging the importance of finance in executing rural electrification projects, it should also be noted that the best way to meet the requirements of projects is by engaging relevant stakeholders from the beginning of the policy process, and this leads us to another pressing challenge for rural electrification in Cameroon.

### 2.4.4  Lack of Grassroots and Other Stakeholder Participation in the Policy-Making Process

Policies are designed to achieve certain goals and these policies depend on the support and cooperation of the target communities to succeed. Policymaking for rural electrification in Cameroon often follows a top-down approach, without involving the local people and other stakeholders. While the policy for rural electrification in Cameroon is made at ministerial department levels, it requires the acceptance and ownership of local communities for projects to succeed. People do not need electricity per se, but they need the services that electricity can provide them, so it is not enough to bring electricity to a community without considering their needs and preferences. Also, through stakeholder engagement, basic requirements for project success can be met. A good example is the level of financing. Local banks are potential partners in funding small-scale electrification schemes for rural areas, but they are not part of the rural electrification efforts of the government. A credit manager said in an interview:

> Many commercial banks in Cameroon do not invest in the rural electrification sector because it is a sector owned and run by the government. Our bank does not know if the sector has been liberalised and only the government invests in that sector. Cameroonian banks have a problem with resources. Financing this kind of project requires long-term investment and

most of the resources of local banks make money from short-term investments…. The level of support the government gives to international banks is more than what we the local banks receive. (Interview with R7, 7th May 2019).

With this, one can see that with the effective involvement of relevant stakeholders, funding and other project resources can be secured. This section has identified and discussed some of the factors that hinder the improvement of rural electricity access in Cameroon. These factors include limited coordination of the rural electrification sector, lack of powers by the REA to make rural electrification policies, lack of funds, lack of grassroots and other stakeholder participation in the policy-making process, insecurity, and reliance on grid extension for rural electrification. These factors have contributed to the low level of rural electrification in Cameroon, which stands at about 25%. The next section will explore some of the possible solutions and best practices that can address these challenges and enhance the rural electrification process in Cameroon.

## 2.4.5   Challenge of Insecurity

Rural electrification in Cameroon is also hampered by the challenge of insecurity. The multiple conflicts in the rural parts of the country are disrupting projects. In the Northern part of the country, there is the Boko Haram insurgency, and in the two English-speaking regions of the North West and South West, there is also an armed conflict.

According to the Presidency and the Ministry of Water Resources and Energy, the priority regions for rural electrification are the East, Far North, North, North East and South West. There is an ongoing crisis in three of these regions and this has affected our projects. When you look at the project we are carrying out for a mini hydro plant in Boa, South West region, construction works were forced to stop due to the insecurity of our collaborators doing the construction. (Interview with R5, 10th May 2022).

These insecurity challenges are also a big problem for project continuity after the conflict. Many contractors have abandoned their projects which are deteriorating. Resuming these projects will incur more costs, requiring more financial allocations (where necessary and possible).

## 2.4.6   Reliance on Grid Extension for Rural Electrification

Electrification policy in Cameroon tends to favor grid extension. The recent creation of the National Electricity Transmission Company (SONATREL) is a strong indication of this preference, and the government intends to continue its efforts to connect more communities to the grid. In the words of staff from an independent power-producing company:

> Here, we produce electricity and are compelled to supply the electricity into the grid where Eneo alone has the right to sell to customers. With this kind of governmental policy, we cannot locate our production facilities in areas that are far from the grid. (Interview with R7, 7th May 2019).

The sale of electricity seems like a monopolistic market in Cameroon and this practice discourages independent power producers from investing in the sector. Cameroon as a country suffers from poor transportation infrastructure and with bad roads, it becomes more difficult to construct transmission lines through areas with bad roads and forest terrain. With the third hydropower potential in Africa, Cameroon has 8 drainage basins. The abundance of rivers offers opportunities for several mini and pico hydro projects all over the country (Fig. 2.10).

Off-grid systems are a viable option with huge prospects for rural areas in Cameroon. Having harnessed only about 1% of its renewable energy potential (excluding hydropower), Cameroon stands to benefit more from exploiting other renewable energy sources it possesses.

## 2.5 Stakeholder Model for Advancement of Rural Electrification in Cameroon

Based on the literature reviewed and primary data collected on rural electrification in Cameroon, it can be seen clearly that the country is blessed with enormous energy resources but suffers from limited rural electrification. While the government of Cameroon, through the REA, seems committed to the rural electrification agenda, several challenges remain:

- Limited co-ordination of the rural electrification sector
- The inability of the REA to make policies
- Lack of funds
- Limited stakeholder participation
- Insecurity
- Reliance on grid extensions.

Based on the challenges culled from the study, the stakeholders below have been identified with their respective roles in Table 2.1 (Fig. 2.11).

The conceptual model developed above is based on the idea of blending stakeholder participation with evidence-based decision-making to achieve effective policymaking and improved rural electrification status. This model, if applied to the rural electrification sector of Cameroon, can help to overcome the existing barriers that have been hindering the progress of rural electrification across the country. In this section, we will review each of the identified challenges and propose solutions based on the conceptual model. We will also assess the feasibility and suitability of the proposed solutions for the Cameroonian context.

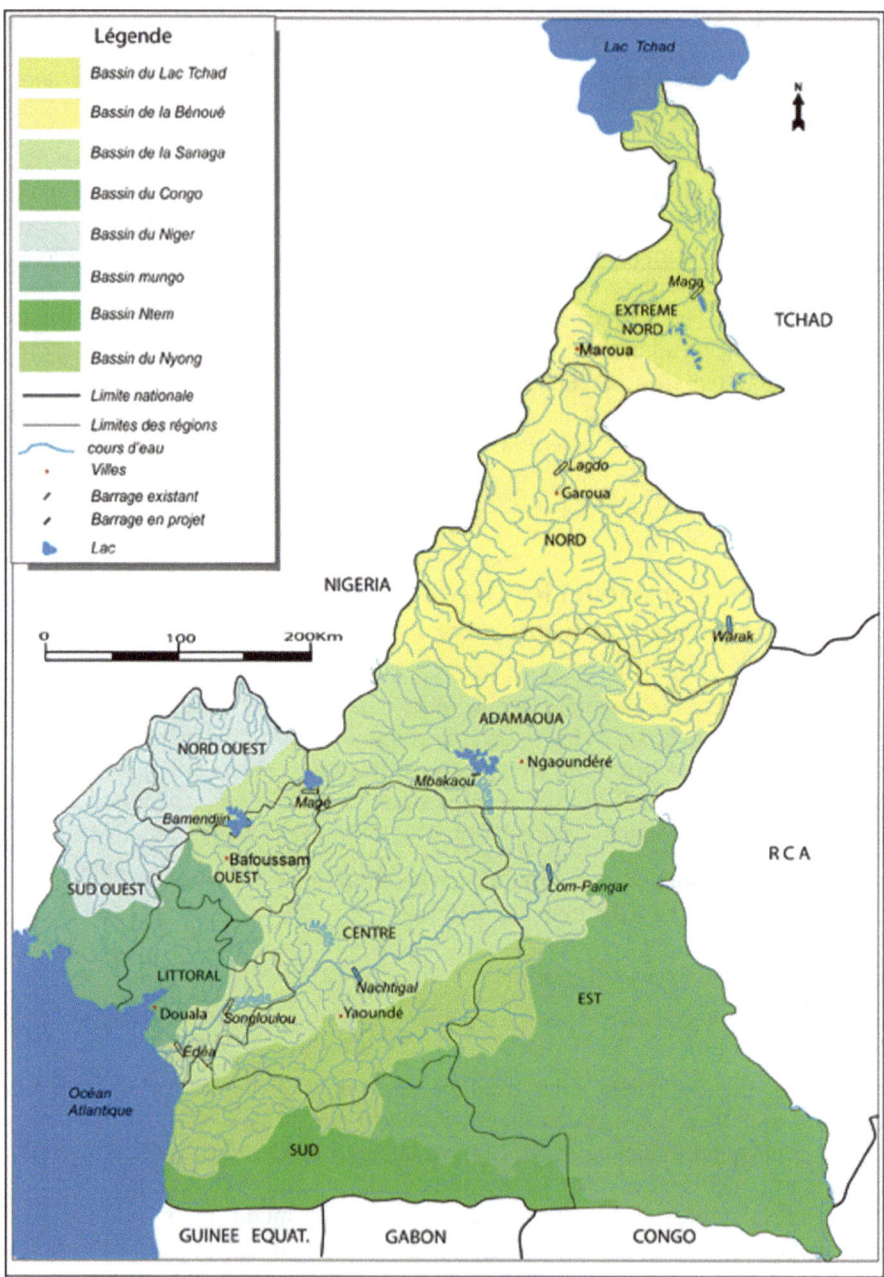

**Fig. 2.10** Drainage and drainage basins (*Source* Cameroon Tour 2009)

**Table 2.1** Table showing challenges, stakeholders, and proposed solutions

| Challenges | Stakeholders | Solutions |
|---|---|---|
| 1. Limited coordination and lack of regulation and appropriate rural electrification policies | • Electricity Sector Regulatory Agency (ESRA) | • Provide regulatory framework<br>• Establish subsidy and tariff schemes |
| | • Ministry of Water Resources and Energy (MWRE) | • Enforce regulations, framework, and policy guidelines |
| | • REA | • Coordinating with relevant stakeholders |
| 2. Inadequate funds | • Banks | • Provide loans to private investors for electrification projects |
| | • Rural Energy Fund | • Provide finance for investments in the sector and also make money available to banks |
| | • Cooperative Unions | • Provide funding |
| | • Private Investors | • Provide funding |
| | • National and International donors | • Provide funding<br>• Provide funding |
| | • Local Governments | • Provide funding |
| 3. The inability of REA to make rural electrification policies | • Office of the President | • Enacting a decree to empower the REA to make rural electrification policies |
| 4. Insecurity | • Central Government | • Suing for ceasefires, dialogue, and addressing existing grievances |
| | • Local Governments | • Organizing community initiatives for dialogue |
| | • NGOs | • Engage in Peace Education and outreach programs |
| 5. Limited Stakeholder participation | • MWRE | • Approving REA policies<br>• Ensuring REA policies adhere to bottom-up policy-making guidelines |
| | • REA | • Ensuring grassroots participation in consultations for rural electrification policy making |
| | • NGOs | • Providing education on the importance of community participation in policy-making<br>• Making information available |

(continued)

**Table 2.1** (continued)

| Challenges | Stakeholders | Solutions |
|---|---|---|
| | • Local Governments | • Providing information on project tenders to the public<br>• Facilitating environmental impact assessments for projects by collecting community surveys and consultative talks |
| 6. Reliance on grid extension | • MWRE | • Design plans for decentralized electrification systems |
| | • REA | • Provide financial and technical support to Local governments, IPPs, and NGOs |
| | Research Centres | • Provide short courses to technicians on how to install and maintain off-grid RETs<br>• Research on RETs<br>• Provides skills and technical knowledge on RETs<br>• Enhance research and available data on renewable energy sites in Cameroon |
| | • Independent Power Producers (IPPs) | • Enhance meteorological data provided by government departments |
| | • Local Governments | • Provide funding for Renewable Energy Projects |
| | • NGOs | • Creating awareness in the local communities about the use of RETs<br>• Creating awareness through community centers |

*Source* Iweh et al. (2023)

The main challenges that we have discussed in the previous section are limited co-ordination of the rural electrification sector, lack of powers by the REA to make rural electrification policies, lack of funds, lack of grassroots and other stakeholder participation in the policy-making process, insecurity, and reliance on grid extension for rural electrification. For each of these challenges, our conceptual model can provide a solution that is based on stakeholder involvement and evidence use.

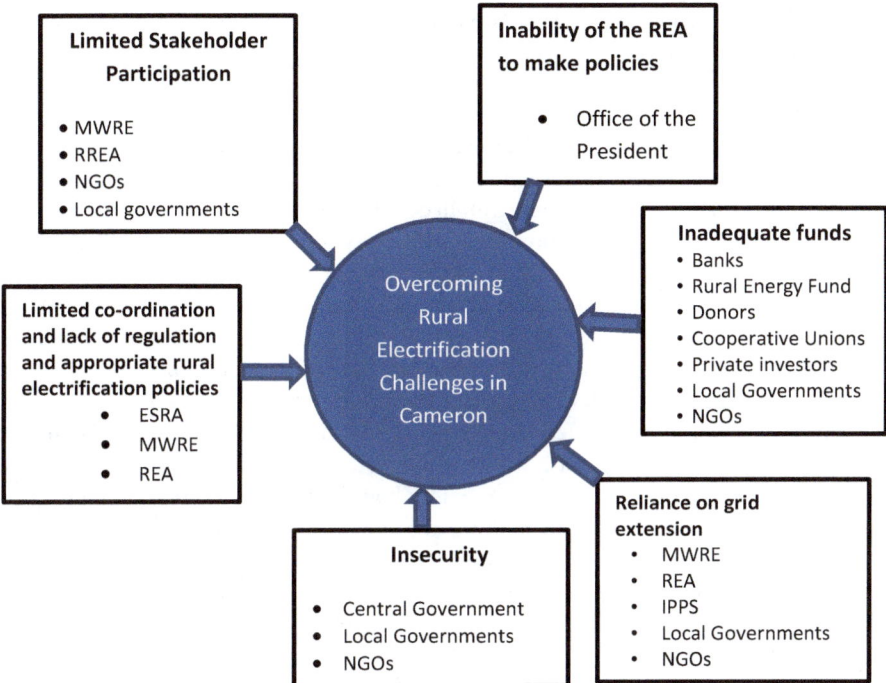

**Fig. 2.11** Conceptual model for overcoming rural electrification challenges in Cameroon

## 2.6 Conclusions

Using a qualitative methodology, the researcher aimed to explore the challenges of rural electrification in Cameroon. The study selected key respondents who are actors in the electricity sector in Cameroon, using the purposive sampling technique. The study found that electricity access has a significant impact on the living standards of rural areas. Electricity brings more opportunities and developmental prospects, as seen globally. However, most rural areas in Cameroon lack electricity and have fewer development opportunities.

Electricity and other energy sources are essential for powering industries and other sectors of the economy in any industrialized society. Cameroon has a vision of becoming an emerging economy by 2035, which requires energy for its achievement. However, the rural electrification master plan aims to achieve 100% electrification of the country by 2035, which raises the question of how emergence can take place without energy to power industries and other sectors of the economy before then. The lack of electricity access needs to be tackled to stimulate economic growth (SDG 8) in rural economies and promote access to clean and modern energy forms for all Cameroonians by 2030 (SDG 7). Falling living standards also relate to the high birth rates in rural areas, which are partly due to no electricity access. At the end of a working day, when rural dwellers return home, the lack of electricity means that they cannot engage in any economic activities at night. As a result, these rural dwellers often resort to other social activities, including sex.

The rural–urban and rural-rural migration patterns in Cameroon have both advantages and disadvantages. While rural areas with electricity and urban areas benefit from youths with skills and potential, this creates a brain drain from rural to urban areas, leading to unbalanced regional development and industrialization across Cameroon. Citizens should be provided with equal opportunities regardless of the part of the country they live in. The lack of government policies to prevent rural–urban migration is a big challenge for food security, as the rural population in Cameroon is responsible for most of the food produced in the country. Women also constitute a large proportion of the rural dwellers engaged in food production.

The link between the lack of electricity access in rural areas and deforestation is evident. Fuel wood is highly demanded by rural dwellers in Cameroon. Their low income levels and lack of electricity limit their energy options in their households. Searching for fuel wood is a task done mainly by women in rural areas and it is quite time-consuming. Time spent on looking for fuel wood could have been invested in other economic ventures if electricity and other energy sources were available and affordable for rural women. With most rural economies in Cameroon being supported by cash crops like cocoa and coffee, these are energy-intensive crops at the level of drying before they can be sold. This process requires the use of locally made ovens with fuel wood as the main energy source. This wood is obtained from forests and therefore contributes to releasing already captured carbon into the atmosphere. Combating climate change (SDG 13) will also be an indirect benefit of rural electrification.

Barriers to rural electrification in Cameroon were identified. It is important to highlight that the Cameroonian electricity sector has been dominated by the government for a long time and seems to have been politicized. With electrification policy being made from the level of government offices, with a politicized bureaucracy in Cameroon, the likelihood of a community being electrified depends on party affiliation and also whether the community has members in government.

Reducing reliance on grid extension to increase access is a very important component. Energy democracy is a very important aspect of economic growth and political power. People can only participate in politics equally when they are economically well-off. Decentralizing electricity generation and empowering rural citizens to be "prosumers" would mean that they also generate revenue while providing themselves with electricity generated locally where it is consumed. Decentralized generation can help to reduce reliance on the grid and also reduce Transmission and Distribution losses.

Attaining the AU Agenda 2063 and the SDGs can be possible with the provision of electricity and other energy sources to all. Energy for all is the backbone of all other existing SDGs and as such one should not underestimate the importance of electricity in the global agenda to eliminate hunger and poverty by the year 2030.

Overall, this study has explored the challenges of rural electrification in Cameroon, using a qualitative methodology and purposive sampling technique. The study has found that rural electrification has a positive impact on the living standards, economic development, and environmental sustainability of rural areas. However, the rural electrification process in Cameroon faces several barriers, such as limited coordination of the sector, lack of powers by the REA to make policies, lack of funds, lack of grassroots and other stakeholder participation in the policy-making process, insecurity, and reliance on grid extension for rural electrification. These barriers have resulted in a low level of rural electrification in Cameroon, which stands at about 25%. The study has also proposed a conceptual model that blends stakeholder participation with evidence-based decision-making to achieve effective policymaking and improved rural electrification status. The study has suggested some solutions and best practices that can address the identified challenges and enhance the rural electrification process in Cameroon, based on the conceptual model.

## 2.6.1 Policy Recommendations

Based on the findings of the study, the following recommendations were made:

The rural electrification policy-making process should be a bottom-up approach. Communities to be electrified should be consulted from the onset before policies are made. Rural communities differ based on economic activities and electricity provided should not just be for lighting but should also add value to the existing economic activity in the area. Otherwise, the presence of electricity for lighting and household use will only increase living costs without necessarily improving the finances of rural dwellers. Existing institutions in the rural electrification sector should be strengthened through laws and mechanisms designed for existing regulations to be enforced.

Decentralized generation should be promoted through the provision of feed-in tariffs for IPPs, subsidies, and tax exemptions for RET equipment being imported into the country.

Information should be made available to potential local funding sources like banks and cooperatives about existing financing schemes for renewable energy projects. Finance should be made available to banks and cooperatives so they can fund local projects from IPPs.

Local banks should see investments in the rural electrification sector as a lucrative venture. Not all electrification projects are expensive. Mini hydro projects are also financially viable and banks can work together in synergy to fund projects by pooling resources together.

Banks should benefit from the technical expertise of the REA, ESRA, MWRE, and NGOs to carry out evaluations of proposed projects.

## 2.6.2　Limitation and Future Research

This study has some limitations that should be acknowledged. First, the study focused on a limited number of respondents who are actors in the electricity sector in Cameroon and did not include the perspectives of rural dwellers who are the beneficiaries of rural electrification. Second, the study relied on qualitative data and did not use any quantitative methods or indicators to measure the impact or effectiveness of rural electrification. Third, the study did not conduct a comparative analysis of rural electrification policies and practices in other countries or regions that could provide useful insights or lessons for Cameroon.

Some avenues for future research are suggested based on these limitations. First, future studies could conduct surveys or interviews with rural dwellers to understand their needs, preferences, and perceptions of rural electrification, and how they use electricity for various purposes. Second, future studies could use quantitative methods or indicators to measure the impact or effectiveness of rural electrification on various aspects of rural development, such as income, education, health, gender equality, etc. Third, future studies could conduct a comparative analysis of rural electrification policies and practices in other countries or regions that have achieved higher levels of rural electrification or have implemented innovative solutions or best practices that could be adapted or replicated in Cameroon

**Acknowledgements**　The author would like to acknowledge and thank Nkumbe Enongene Rex for the earlier version of this work. This chapter builds on the project he accomplished at PAUWES, University of Tlemcen, Algeria, and financed by PAUWES.

# References

AER (2015) Rural Electrification Agency. Retrieved from
AfDB (2009a) Project to strengthen and extend the electricity transmission and distribution networks. The Republic of Cameroon. Accessed on 6th April 2019 from https://www.afdb.org/fileadmin/uploads/afdb/Documents/Project-and-Operations/-%20Cameroon%20-%20AR%20Electricity%20Project%20-%5B1%5D.pdf
AfDB (2009b). Cameroon: rural electrification project. Retrieved from
AfDB (2021) Country priority plan and diagnostic of the electricity sector Cameroon. https://www.afdb.org/sites/default/files/2021/11/22/cameroon.pdf
Agence de l'Electrification Rurale du Cameroun (AER) (2015) Missions of AER. Accessed 25th August 2018 from http://www.aer.cm/menu/missions-of-aer
ARE (2015) Hybrid mini-grids for rural electrification: lessons learned. The alliance for rural electrification
Brohman J (1996) Popular development: rethinking the theory and practice of development. Blackwell Publishers, Oxford
Business in Cameroon (2018) Electricity: Cameroon's installed capacity stands at about 1.442 MW as of June end. https://www.businessincameroon.com/electricity/1807-8309-electricity-cameroon-s-installed-capacity-stands-at-about-1-442mw-as-at-june-end

Buzanakova M (2014) Energy sector development in Cameroon: challenges and opportunities. Retrieved from https://www.researchgate.net/publication/269695637_Energy_Sector_Develo pment_in_Cameroon_Challenges_and_Opportunities

Buzanakova A (2014) Cameroon natural gas reserves 1980–2013. Accessed on 6th December 2018, from http://cameroon.opendataforafrica.org/qowhahe/cameroon-natural-gas-reserves-1980-2013.

Cairn International (2016) The impact of rural electrification: challenges and ways forward. Retrieved from https://www.cairn-int.info/article-E_EDD_293_0055--the-impact-of-rural-ele ctrification.htm

Cameroon Tour (2009) Drainage and drainage basins in Cameroon. Accessed on 27th May 2019 from https://cameroon-tour.com/geography/drainage.html

Castalia (2015) Evaluation of rural electrification concessions in sub-Saharan Africa: detailed case study: Cameroon. World Bank Group

Chaurey A, Kandpal TC (2010) Assessment and evaluation of PV based decentralized rural electrification: an overview. Renew Sustain Energy Rev 14(2266–2278)

Conyers D (1985) Rural regional planning: towards an operational theory. Prog Plan 23:3–66

Cornia GA, Jolly R, Stewart F (1987a) Adjustment with a human face. Oxford University Press

Cornia G, Jolly R, Stewart F (1987b) Adjustment with a human face: protecting the vulnerable and protecting growth. Oxford University Press, Oxford

Crousillat E, Hamilton R, Antmann P (2010) Addressing the electricity access gap. Background paper for the world bank group energy sector strategy. Retrieved from http://siteresources.worldb ank.org/EXTESC/Resources/Addressing_the_Electricity_Access_Gap.pdf. Accessed 18 Aug 2018.

Decree 99/193 dated 8 September 1999

Decree no. 2013/204 of 28 June 2013

Decree no.2006/406 dated 29 November 2006

Decree of 1999 (Decree no. 99/125 dated 15 June 1999

Deloitte (2023). Solutions for rural electrification in developing countries

Djoedjom CF, Zhao X (2018) Current status of renewable energy in Cameroon. North Am Acad Res 1(2):71–80

Fenster T (1993) Conceptualizing community planning within a multicultural context. Town Plan Rev

Fenster T (1993b) Settlement planning and participation under principles of pluralism. Prog Plan 39:171–242

Fleming P (1991) Community participation in rural development. In: Rural development theory and practice. Routledge

Fleming S (1991) Between the household: researching community organization and networks. IDS Bulletin 22

Fotsing I, Njomo D, Tchinda R (2014) Analysis of demand and supply of electrical energy in Cameroon: Influence of meteorological parameters on the monthly power peak of south and north interconnected electricity networks. Int J Energy Power Eng 3(4):168–185

Fowler A (1991) The role of NGOs in changing state-society relations: perspectives from Eastern and Southern Africa. Dev Policy Rev 9:53–84

FUSS, Cameroon, and renewable energy. country at a glance. Fed Univ Appl Sci 2013:1–2. https:// www.laurea.fi/en/document/Documents/CameroonFactSheet.pdf. Accessed 15 Aug 2022

Fuss S (2013) Renewable energy development plan for Cameroon: a feasibility study based on cost optimization modeling using OSeMOSYS (Master thesis). KTH Royal Institute of Technology

Guefano S, Tamba JG, Monkam L, Bonoma B (2020) Forecast for Cameroon's residential electricity demand based on the multilinear regression model. Energy Power Eng 12(6):182–192

Hisham Z (2010) Rural electrification programs: a comparative study on Malaysia's success story. Energy policy research group working paper No. 1019

How We Made It In Africa (2017) Rural electrification in Africa: an economic development opportunity. Retrieved from https://www.howwemadeitinafrica.com/rural-electrification-africa-economic-development-opportunity/58403/

IEA (2014a) Africa energy outlook: a focus on energy prospects in sub-Saharan Africa. https://www.iea.org/reports/africa-energy-outlook-2014

IEA (2014b) Africa energy outlook: a focus on energy prospects in sub-Saharan Africa. https://www.iea.org/reports/africa-energy-outlook-2014

IEA (2014c) African energy outlook: a focus on energy prospects in Sub-Saharan Africa. In: World energy outlook special report. Author, Paris

IEA (2014d) Africa energy outlook. Retrieved from https://www.iea.org/reports/africa-energy-outlook-2014

IEA (2016) Electricity generation by fuel: Cameroon 1990–2016. Accessed on 4th April 2021, from https://www.iea.org/statistics/?ountry=CAMEROON&year=2016&category=Electricity&indicator=ElecGenByFuel&mode=chart&dataTable=ELECTRICITYANDHEAT.

IEA (2017) Energy access outlook: from poverty to prosperity. World energy outlook special report. Author, Paris

IEA (2019) Africa energy outlook 2019: world energy outlook special report. Author, Paris

IsDB (2017a) International call for tender limited to IDB member countries for contractor selection: Cameroon. Accessed on 6th April 2019 from https://www.isdb.org/tenders/international-call-for-tender-limied-to-idb-member-countries-for-contractor-selection

IsDB (2017b) Cameroon: Rural Electrification Project Phase II. Retrieved from

Iweh CD, Ayuketah YJ, Gyamfi S, Tanyi E, Effah-Donyina E, Diawuo FA (2023) Driving the clean energy transition in Cameroon: a sustainable pathway to meet the Paris climate accord and the power supply/demand gap. Front Sustain Cities 5(1062482)

Javadi FS, Rismanchi B, Sarraf M, Afshar O, Saidur R, Ping HW, Rahim NA (2013) Global policy of rural electrification. Renew Sustain Energy Rev 19(402–416)

Jepsen AL, Eskerod P (2009) Stakeholder analysis in projects: challenges in using current guidelines in the real world. Int J Project Manage 27:335–343

Karl M (2002a) Measuring the immeasurable: Planning monitoring & evaluation of empowerment processes. UNDP

Karl M (2002b) Participatory policy reform from a sustainable livelihoods perspective: review of concepts and practical experiences. LSP working paper 3, participation, policy, and local governance sub-programme. FAO, Rome. http://www.fao.org/docrep/006/ad688e/ad688e03.htm

Law no. 99/016 of 22 December 1999

Lekunze N (2001a) The challenges of indigenous peoples: example of Cameroon. International Labour Office

Lekunze RN (2001b) Assessing stakeholder participation in integrated water resource management: the role of youth in community water management projects in Cameroon. M.Sc. thesis. Lund University, Lund

Liang X, Yu T, Guo L (2017) Understanding stakeholders' influence on project success with a new SNA method: a case study of the green retrofit in China. Sustainability 9(10):1927

Lighting Africa (2012) Policy report note: Cameroon. https://www.lightingafrica.org/wp-content/uploads/2013/12/Cameroon_Policy_Report_Note.pd.

McElroy B, Mills C (2000) Managing stakeholders. In: Turner RJ, Simister SJ (eds) Gower handbook of project management. Gower, Aldershot

Midgley J (1986) Community participation, social development, and the state. Methuen, London

Ministère de l'Energie et de l'Eau (MINEE) «Présentation de AER» http://www.minee.cm/index.php?page=aer

Moser C (1989) Gender planning in the third world: meeting practical and strategic gender need. World Dev 17:1799–1825

Niez A (2010) Comparative study on rural electrification policies in emerging economies: keys to successful policies. International Energy Agency

Njomo D, Njomo M, Njomo L (2021) Rural electrification in Cameroon: a review of challenges and opportunities. Int J Energy Econ Policy 11(1):412–422

Nkongho RN, Tchinda R, Njomo D (2002) Electricity sector reform in Cameroon: Is privatization the solution? Energy Policy 30(11–12):999–1009

Onyeji I, Bazilian M, Nussbaumer P (2012) Contextualizing electricity access in Sub-Saharan Africa. Energy Sustain Dev 16:520–527

Paul S (1986) Community participation in development projects: the World Bank experience. In: World Bank discussion papers, 6. Washington, DC

Qmerit (2022) The top 5 rural electrification challenges

REA (2015) Rural electrification agency. Retrieved from http://www.rea.cm/

REEP (2013) Renewable energy and energy efficiency programme. Retrieved from http://www.reep-cameroon.org/

Rehman A, Deyuan Z (2018) Investigating the linkage between economic growth, electricity access, energy use, and population growth in Pakistan. MDPI J Appl Sci 8:2442. https://doi.org/10.3390/app8122442

Renewable Energy and Energy Efficiency Partnership (REEP) (2013) Cameroon (2012): existence of an energy framework and programs to promote sustainable energy. Accessed 25th Aug 2018 from https://www.reeep.org/cameroon-2012

Rietbergen-McCracken J (2011a). Participatory policymaking. World Bank

Rietbergen-McCracken J (2011b) Participatory development planning, working paper, world alliance for citizen participation (CIVICUS), participative governance project, PG Exchange initiative. http://www.pgexchange.org/images/toolkits/PGX_F_Participatory%20Development%20Planning.pdf

Rural Electrification Master Plan (2016) Report No: PAD2677. The World Bank. Retrieved from https://documents1.worldbank.org/curated/en/260941545015657290/pdf/Cameroon-Rural-Electricity-Access-Project-for-Underserved-Regions-Project.pdf

Suhlrie L, Bartram J, Burns J, Joca L, Tomaro J, Rehfuess E (2018) The role of energy in health facilities: a conceptual framework and complementary data assessment in Malawi. PLoS ONE 13(7):e0200261. https://doi.org/10.1371/journal.pone.0200261

Sutcliffe S, Court J (2006) What is it? How does it work? What is relevance for developing countries? Overseas Development Institute. Accessed on 13th July 2019, from https://www.odi.org/sites/odi.org.uk/files/odi-assets/publications-opinion-files/3683.pdf

Too EG, Weaver P (2014) The management of project management: a conceptual framework for project governance. Int J Project Manage 32(8):1382–1394

Tsoukiàs A, Montibeller G, Lucertini G, Belton V (2013) Policy analytics: an agenda for research and practice. Euro J Decis Process 1(1–2):115–134. https://doi.org/10.1007/s40070-013-0008-3

UN General Assembly (2015) Transforming our world: the 2030 agenda for sustainable development, 21st October, A/RES/70/1. Available at http://www.refworld.org/docid/57b6e3e44.html. Accessed 18 Aug 2018.

UNDP (2012) Africa human development report 2012: towards a food secure future. Retrieved from http://hdr.undp.org/en/content/africa-human-development-report-2012-towards-food-secure-future

UNIDO (2016) Hydropower in Cameroon. United Nations Industrial Development Organ 2016. Accessed on 11th Aug 2016, from http://www.unido.it/eng/idro.php

United Nations Center for Human Settlements (UNCHS) (1984) Community participation in the execution of low-income housing projects. Nairobi, Kenya

United Nations Environment Programme (UNEP) (2001) Freshwater: a global crisis of water security and basic water provision. Towards earth summit 2002, environment briefing Paper No. 1

Weible CM, Carter DP (2017) Advancing policy process research at its overlap with public management scholarship and nonprofit and voluntary action studies. Policy Stud J 45(1):22–49

Wilson JQ (1995) Political Organisations. Princeton University Press, Princeton, NJ

World Bank Group (2018) Cameroon. Accessed on 20th June 2019 from https://data.worldbank.org/country/Cameroon

World Bank (2019) Electrification efforts in Sub-Saharan Africa must address root causes of low access. Retrieved from https://www.worldbank.org/en/region/afr/publication/electricity-access-sub-saharan-af

Young E, Quinn L (2012) Making research evidence matter: a guide to policy advocacy in transition countries. Open Society Foundations, Budapest

# Chapter 3
# Exploring the Energy Transition to LPG in the Nigerian Household Sector: A Scenario-Based Modeling Approach

**Abstract** Scenario-based models based on plausible and acceptable descriptions of the future have time and again been employed to study the effect of an evolving energy system. This study employed scenario-based modeling using the Long-Range Energy Alternatives Planning System (LEAP) to explore the dynamics of the energy transition to cleaner energy in the Nigerian household sector. Our analysis showed that the household sector has and will continue to contribute the largest share to the overall energy demand in Nigeria within the period under study (2010–2030). To achieve sustainable development and eradicate energy poverty within the Nigerian household sector an alternative energy transition scenario was explored. This scenario identified bottlenecks that could hinder the mainstreaming of Liquefied Petroleum Gas (LPG) to substitute traditional fuelwood in the household sector and modeled a possible future where these bottlenecks are tackled, and LPG becomes the fuel of choice within the household sector in Nigeria.

*What is in for the readers of this chapter?* In this Chapter, researchers can avail the scenario-based modeling method to explore the dynamics of the energy transition to cleaner energy in Nigeria and other countries. Policymakers can use the results and recommendations of the scenario analysis to formulate and evaluate policies and strategies for promoting liquefied petroleum gas as a substitute for traditional fuelwood in the household sector. Practitioners can learn from the bottlenecks and solutions for mainstreaming liquefied petroleum gas in the household sector and apply them in their contexts. University's research libraries and research centers can benefit from the data and information on the energy demand and supply in the Nigerian household sector and the potential of liquefied petroleum gas to reduce greenhouse gas emissions and improve health outcomes.

© The Author(s), under exclusive license to Springer Nature Switzerland AG 2024
H. Qudrat-Ullah, *Exploring the Dynamics of Renewable Energy and Sustainable Development in Africa*, Advances in African Economic, Social and Political Development, https://doi.org/10.1007/978-3-031-48528-2_3

## 3.1  Introduction

Energy is essential for sustaining the well-being of a society, as it enables access to education, health, and quality of life (Ashagidigbi et al. 2020a, b). Moreover, the economic development of nations over time has been driven by fundamental changes in the energy mix. One of the main motivations for global energy transitions is the security of the energy supply (Su et al. 2019a, b). The United Nations' Sustainable Development Goal No. 7 explicitly calls for "access to affordable, reliable, sustainable and modern energy for all" (Dioha and Kumar 2020a, b). Therefore, renewable energy (RE) sources are essential for the household, commercial, and industrial sectors of the economy. They can provide clean, sustainable, and affordable energy for people around the world. That is why many energy policymakers are aiming to develop and use RE sources as their main goal.

The energy transition is the gradual change that makes the energy system more diverse over time (Yergin et al. 2013a, b). Two main reasons have prompted the need for a new global energy transition in this century. The first reason is the worry about how fossil fuels, which are the main source of energy now, affect the climate. The second reason is the rise of developing countries as the new economic powers and the growing global energy demand from these countries. Many organizations, such as the World Bank, the United Nations Development Programme, and the World Energy Council, are trying to reduce energy poverty, especially in households, as part of their efforts to support the energy transition (Pachauri and Spreng 2011).

The energy transition that is happening now is different from the past ones in many ways. The main reason for this transition is to deal with climate change and how it affects people and nature. The need for low-carbon energy is based on the science from the Intergovernmental Panel on Climate Change (IPCC), which is the UN organization that studies climate change (WEF 2020). Second, it involves not only a shift in energy sources but also a shift in energy consumption patterns and behaviors. This requires changes in infrastructure, institutions, policies, markets, and social norms that enable a more efficient and sustainable use of energy (TWI 2021). Third, it is influenced by multiple actors and factors at different levels and scales, such as governments, businesses, civil society, consumers, geopolitics, innovation, and finance. This makes the energy transition a complex and dynamic process that varies across regions and countries depending on their contexts and capacities (WEF 2021).

In Nigeria, one of Africa's most populous and fastest-growing economies, household energy demand accounts for about 20% of total final energy consumption (International Energy Agency [IEA] 2019). However, most households rely on traditional biomass (such as fuelwood) for cooking and water heating, which poses health risks for humans and environmental risks such as deforestation in the long run if not curbed. Therefore, there is an urgent need to promote clean fuel substitution in Nigeria's household sector.

Clean fuel substitution refers to the process of replacing traditional biomass fuels with cleaner alternatives such as kerosene, liquefied petroleum gas (LPG), or electricity for cooking and other domestic purposes. Clean fuel substitution can bring multiple benefits for households, such as improved indoor air quality, reduced drudgery and time spent on fuel collection, increased convenience and comfort, and lower greenhouse gas emissions (Oyedepo 2012). However, clean fuel substitution is not a straightforward process and depends on various factors such as availability, affordability, accessibility, acceptability, and awareness of different fuel options (Schlag and Zuzarte 2008).

According to a recent study by Heinrich Böll Stiftung (2021), the predominant energy resources for domestic and commercial uses in Nigeria are fuelwood (41%), charcoal (1%), kerosene (53%), LPG (4.5%), and electricity (0.5%). The study also found that income levels, poverty rates, urbanization rates, cultural preferences, and policy interventions are some of the key drivers for clean fuel substitution in Nigeria. The study recommended that to expand the demand for clean cooking in Nigeria, there is a need to adopt a multi-stakeholder approach that involves government agencies, private sector actors, civil society organizations, and local communities in designing and implementing effective policies and programs that address the barriers and opportunities for clean fuel substitution.

The main research question that this study addresses is: What are the possible energy transition pathways for Nigeria's household sector to meet its clean energy access goals? To answer this question, we develop an integrated system model that captures both thermal and non-thermal uses of energy in Nigeria's household sector using Long-range Energy Alternatives Planning System (LEAP) software. We examine future energy demand dynamics under various economic conditions and clean fuel substitution scenarios for different energy transition pathways. We also examine the socio-economic implications of switching to clean fuel in Nigeria's household sector.

This study contributes to the literature on energy transition and sustainable development in several ways. First, it provides an innovative application of an integrated LEAP-based model that simultaneously accounts for major economic sectors, sub-sectors, end-users, and devices that consume energy. Second, it designs and assesses liquefied petroleum gas (LPG) substitution scenarios in Nigeria's household sector using case-specific data and assumptions. Third, it evaluates the impacts of clean fuel switching on energy security, greenhouse gas emissions, health outcomes, and household expenditure.

This study is novel and innovative in at least two aspects. First, it applies an integrated LEAP-based model that simultaneously accounts for major economic sectors, sub-sectors, end-users, and devices that consume energy in Nigeria's household sector. This allows for a comprehensive and realistic analysis of energy demand dynamics and clean fuel substitution scenarios. Second, it designs and assesses LPG substitution scenarios using case-specific data and assumptions that reflect the current and future conditions of Nigeria's household sector. This provides valuable insights for planning and decision-making to achieve clean energy access goals. To the best of our knowledge, this is the first study that adopts such an integrated and case-specific approach to examine energy transition pathways for Nigeria's household sector.

The rest of the chapter is organized as follows. Section 3.2 provides an overview of energy transition and sustainable development, socio-economic and environmental dynamics of energy use in Sub-Saharan Africa (SSA), and a review of Nigeria's household sector. Section 3.3 describes the development of the demand model and scenarios. Section 3.4 presents and discusses results for various scenarios of energy demand and LPG use. Section 3.5 concludes.

## 3.2 Energy Transition and Sustainable Development

Achieving sustainable development requires a shift from fossil and $CO_2$-emitting fuels to cleaner energy sources (Li et al. 2020a, b; Nsafon et al. 2023). In Africa, this transition faces a unique challenge due to the coexistence of traditional and modern energy systems and practices, which affects the provision of modern energy services for communities that lack them. Energy access in this context can be understood as access to electricity and access to modern fuels for cooking and heating that replace traditional biomass. The Advisory Group on Energy and Climate Change (AGECC) proposes three tiers of energy access as shown (Fig. 3.1) (AGECC 2016). Energy transition to modern energy uses is crucial for the sustainable development of a country, as it reflects the country's level of development, as shown by the energy ladder (van der Mandelli et al. 2016a, b). In Africa, people who are at the "bottom" of the ladder use wood/biomass for cooking and kerosene for lighting, and they gradually move up the ladder as more development activities are realized. Policy frameworks and regulations are needed to accelerate this transition to modern electricity sources for rural and urban-poor Africans.

However, the transition to cleaner energy is not only about decarbonization, but also about ensuring social justice, economic prosperity, and environmental protection for all (IRENA 2021; Nsafon et al. 2023). African countries have an opportunity to

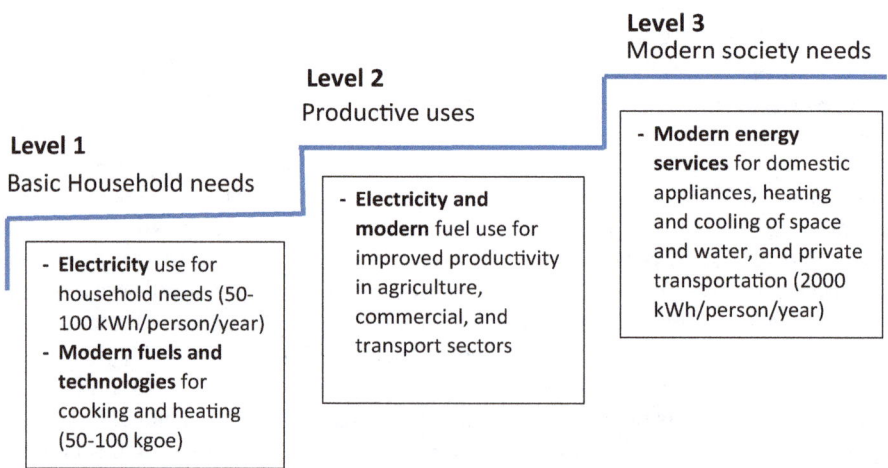

**Fig. 3.1** Incremental levels of access to energy services. *Source* AGECC (2010)

leapfrog fossil fuel technologies and adopt a more sustainable, climate-friendly power strategy aligned with the Paris Agreement and low-carbon growth (IRENA 2021; Müller et al. 2021). However, this also requires addressing the multiple challenges of energy security, economic growth, and affordable access that the continent faces (Oyedepo et al. 2012; Worlddata.info n.d.). Therefore, an incremental transition that emphasizes low-carbon development and takes into account the local context and needs is the most feasible and pragmatic approach to transform the region's economy and address climate change challenges.

### 3.2.1 Energy Use and Its Impact on the Economy and Environment of Sub-Saharan Africa

Sub-Saharan Africa (SSA) is a region that faces many challenges in terms of energy access, economic development, and environmental sustainability. The household sector, which accounts for about 65% of the total final energy consumption in SSA, is mainly dependent on solid fuels such as fuelwood, charcoal, and dung for cooking and heating purposes (Hauff et al. 2019). These fuels are not only inefficient and costly, but also have negative impacts on human health and the environment, such as indoor air pollution, deforestation, and greenhouse gas emissions (Sepp et al. 2014a, b).

In contrast, the global average share of the household sector's energy consumption is only 22%, and less than 20% in advanced economies (Hauff et al. 2019). Moreover, other developing regions outside Africa have witnessed a decline in the use of solid fuels from 1990 to 2010, while SSA has experienced an increase of 2.6% annually during the same period (Sunil and Govinda 2014a, b, c). This indicates that SSA is lagging in the transition to cleaner and more modern forms of energy, such as liquefied petroleum gas (LPG), biogas, solar, and electricity.

The COVID-19 pandemic has further exposed the vulnerability of SSA's energy systems and the importance of reliable and affordable energy access for economic recovery and resilience. Electricity is essential for enabling remote work, online education, digital governance, and health services, as well as providing clean water for hygiene purposes. However, according to the World Bank (2019), about 60% of SSA's health facilities lack access to electricity, and only 43% of the population has access to grid electricity. Furthermore, electricity supply is often unreliable, expensive, and dependent on fossil fuels, which contribute to climate change and air pollution.

Therefore, there is an urgent need for SSA to adopt a more sustainable and inclusive energy policy that can address the multiple dimensions of energy poverty, economic growth, and environmental protection. Some of the potential benefits of such a policy include:

- Creating more jobs and income opportunities in the RE sector, which can stimulate economic recovery and diversification (WRI 2021).

- Improving health outcomes and quality of life for millions of people who suffer from respiratory diseases and premature deaths due to exposure to indoor air pollution from solid fuels (Sepp et al. 2014a, b).
- Reducing greenhouse gas emissions and enhancing climate change mitigation and adaptation efforts by increasing the share of low-carbon energy sources in the energy mix (Bekun et al. 2021).
- Enhancing energy security and resilience by reducing dependence on imported fossil fuels and increasing access to decentralized and off-grid RE solutions (Oxfam 2017).

To achieve these benefits, SSA will need to overcome several barriers and challenges that hinder the development and deployment of clean energy technologies and services. Some of these include:

- Lack of adequate financing and investment for RE projects, especially in rural areas where the demand is high but the returns are low (Oxfam 2017).
- Lack of supportive policy frameworks and regulatory environments that can create incentives and reduce risks for private sector involvement in RE markets (OECD 2019).
- Lack of technical capacity and skills among local actors, such as entrepreneurs, technicians, consumers, and policymakers, to design, install, operate, and maintain RE systems (Oxfam 2017).
- Lack of awareness and social acceptance of RE options among potential users, especially women who are often the primary decision-makers and users of household energy (OECD 2019).

To address these barriers and challenges, SSA will need to adopt a holistic and participatory approach that can leverage the strengths and opportunities of various stakeholders, such as governments, donors, civil society organizations, private sector actors, communities, and households. Some of the key elements of such an approach include:

- Mobilizing domestic and international resources to finance RE projects that are aligned with national development priorities and local needs (WRI 2021).
- Developing and implementing policies and regulations that can foster a conducive environment for RE development, such as feed-in tariffs, subsidies, tax incentives, standards, and quality assurance mechanisms (OECD 2019).
- Building technical capacity and skills among local actors through training, education, knowledge sharing, and technology transfer programs (Oxfam 2017).
- Promoting awareness and social acceptance of RE options among potential users through information campaigns, demonstrations, participatory planning, and gender-sensitive approaches (OECD 2019).

Based on this analysis, we can see SSA faces a complex and multifaceted energy challenge that requires a comprehensive and integrated energy policy that can balance the economic, social, and environmental dimensions of sustainable development. RE can play a vital role in achieving this balance, but it will require concerted efforts from various stakeholders to overcome the existing barriers and challenges. By doing so, SSA can not only recover from the COVID-19 crisis but also transform its energy systems for a more prosperous and resilient future.

Figure 3.2 provides the data of the top countries in each sub-region of sub-Saharan Africa that rely on traditional biomass. As the figure illustrates, a sizeable urban population still relies on traditional means of cooking across all sub-regions in sub-Saharan Africa.

The economic impacts of reliance on traditional sources of energy are enormous. In the year 2010, the total annual expenditure on traditional fuels for cooking (mainly charcoal and fuelwood) in the sub-Saharan region was estimated at around US$ 12 billion. To put things in perspective, this is around 1% of the region's 2010 GDP-it is projected to have more than doubled by 2020 to become around US$29 million (Tsan et al. 2014).

Global warming associated with greenhouse gases and energy constraints are two key threats to sustainable economic development (Maji 2019; Adams and Acheampon 2019). Concerning the environmental consequence of the production and use of solid fuels for cooking, studies have revealed that more than 300 million tonnes of wood are consumed yearly across the region of sub-Saharan Africa (Tsan et al. 2014). This has the consequence of degradation of forest reserves and loss of biodiversity. Furthermore, about 120–380 Mt $CO_2e$ of Kyoto Protocol greenhouse gases (and around 600 Mt $CO_2e$ when particulate matter is added) are generated in the region as a direct result of the use of solid fuel and charcoal production (Lambe et al. 2015). Thus, the transition to cleaner use of energy in the household sector becomes an urgent necessity that policymakers have to address.

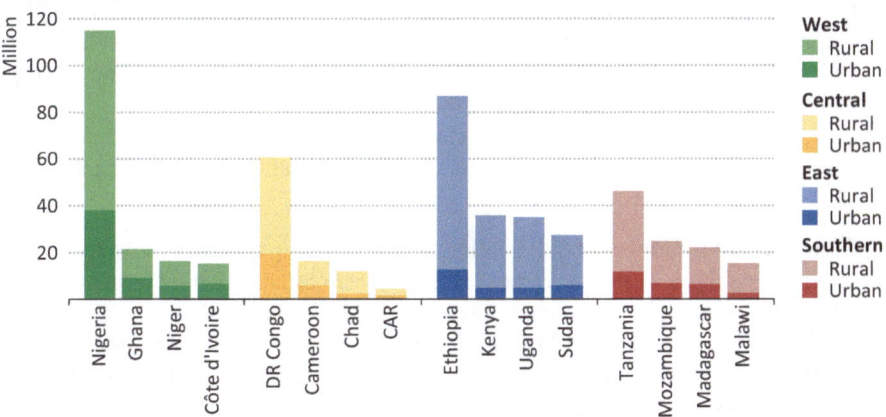

**Fig. 3.2** Largest populations cooking with traditional biomass in SSA. *Source* (Lambe et al. 2015)

## 3.2.2  Energy Transition and Energy Demand in the Nigerian Household Sector

Household energy transition refers to the process of changing the type and quality of energy sources used by households for different purposes, such as cooking, heating, lighting, and appliances. Household energy transition is an important aspect of sustainable development, as it affects the welfare, health, and environmental impacts of household energy consumption (Kowsari and Zerriffi 2011).

One of the key questions that researchers and policymakers have tried to answer is what factors influence the household energy transition and how they interact with each other. This is particularly relevant for developing countries, where many households still rely on traditional biomass fuels, such as fuelwood, charcoal, and dung, which are inefficient, costly, and harmful to human health and the environment (Sepp et al. 2014a, b).

Among the various factors that affect household energy transition, income is often considered the most important one. The energy ladder theory suggests that as household income increases, households tend to switch from traditional biomass fuels to more modern and cleaner fuels, such as kerosene, liquefied petroleum gas (LPG), biogas, solar, and electricity (Leach 1992). This implies that household energy transition is a function of economic development and poverty reduction.

However, income alone cannot fully explain the household energy transition, as other factors also play a role, such as:

- Prices and availability of different energy sources, affect the affordability and accessibility of household energy options (Faiella et al. 2022).
- Preferences and attitudes of household members, especially women who are often the main decision-makers and users of household energy (OECD 2019).
- Cultural and social norms, influence the acceptance and adoption of new energy technologies and practices (Sunil and Govinda 2014a, b, c).
- Policies and regulations, which create incentives or barriers for household energy transition through subsidies, taxes, standards, and quality assurance mechanisms (Bekun et al. 2021).

Moreover, household energy transition is not only influenced by these factors but also influences them in return. For example, household energy transition can affect the demand and supply of different energy sources in the market, which can then affect their prices and availability. Household energy transition can also affect the preferences and attitudes of household members towards different energy options, as they gain more experience and information about them. Household energy transition can also affect the cultural and social norms that shape the energy behavior of households, as they adopt new practices and technologies that may differ from their traditional ones. Household energy transition can also affect the policies and regulations that govern the energy sector, as they create new opportunities and challenges for policymakers and regulators.

Therefore, household energy transition is a complex and dynamic phenomenon that depends on multiple factors that interact with each other in different contexts and over time. Understanding these factors and their interactions can help design more effective and inclusive policies and interventions that can support household energy transition and contribute to sustainable development goals.

One of the major challenges for sustainable development is the rapid population growth, especially in developing countries like Nigeria. Population growth affects the environment and the economy in various ways, such as increasing the demand for resources, generating more waste and pollution, creating pressure on infrastructure and services, and reducing the income per capita and the quality of life (Engelman 2009; Grossman 2012). Other empirical studies assert that the scale of human populations is a principal driving force, along with consumption and technology, of threats to the environment and more generally to sustainability (Knight and Rosa 2011a, b). Nigeria has one of the highest population growth rates in the world, with an average of 2.6% per year from 2011 to 2020 (World Bank 2021), which translates to an additional 5 million people per year, given an average number of 5.0 persons per household. Table 3.1 and Fig. 3.3 show the growth in the population and number of households in Nigeria respectively from 2010 to 2017. This rapid population growth poses a serious threat to Nigeria's sustainable development goals, as it outstrips the economic growth and social development of the country (Victoria-Foye et al. 2020). Therefore, there is an urgent need for Nigeria to adopt effective population policies and strategies that can balance the population size with the available resources and opportunities.

Figure 3.4 shows the distribution of households based on the choice of fuel for cooking from the year 2010 to the year 2021. An average of 72.01% of households in Nigeria used fuelwood to cook. Households that used kerosene largely saw a decline from about 23.80% in 2010 to 9.00% by 2017. Households that used LPG more than doubled from about 3.13% in 2010 to 19.46% by 2021. All other fuels remained marginal, about 3.11% from 2010 to 2021.

In Nigeria, traditional fuel accounted for about 93.67% in 2010 as recorded in Table 3.2.

Dioha and Emodi (2019) report that Nigeria used 9.6 Mtoe of electricity in 2018 and estimate that this will increase to 37.5 Mtoe by 2040 in the STEPS scenario and to 59.3 Mtoe in the Africa Case scenario. The residential sector uses the most electricity, followed by productive uses and transport (Dioha and Emodi 2019). The STEPS scenario reflects the current policies and plans of the Nigerian government, while the Africa Case scenario assumes a more ambitious development of the energy

**Table 3.1** Nigerian population (million people)

| Year | 2013 | 2014 | 2015 | 2016 | 2017 | 2018 | 2019 | 2020 | 2021 | 2022 | 2023 |
|---|---|---|---|---|---|---|---|---|---|---|---|
| Population | 171.8 | 176.4 | 181.1 | 186.0 | 190.9 | 195.9 | 201.0 | 206.1 | 195.9 | 201.0 | 211.4 |

*Source* World population review (n.d.). Nigeria population 2023 (Live). Retrieved from https://worldpopulationreview.com/countries/nigeria-population

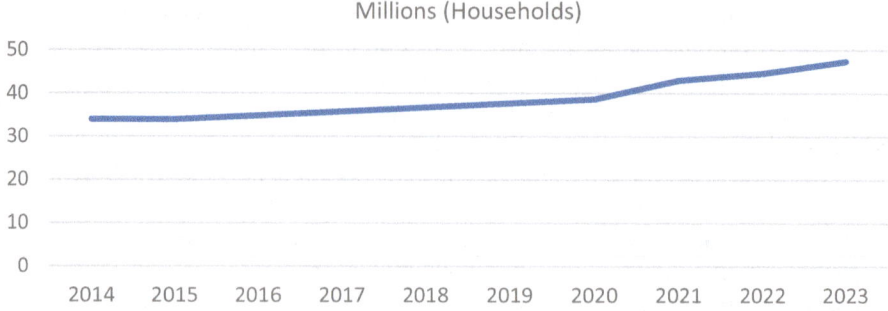

**Fig. 3.3**  Number of households in Nigeria (2014–2023). *Source* World Population Prospects 2019 Volume I: Comprehensive tables, United Nations Department of Economic and Social Affairs; World Population Review. (n.d.). Nigeria population 2023 (live). Retrieved July 18, 2023, from https://worldpopulationreview.com/countries/nigeria-population

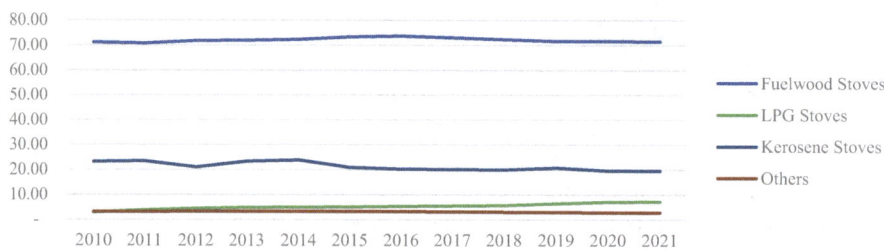

**Fig. 3.4**  Dynamics of fuel used for cooking and water heating (2010–2021) (*Source* CBN (2015) and NBS (2010–2021)

**Table 3.2**  Energy demand in the household sector for the base year 2010

| End-use | Fossil fuels | | Electricity | | Traditional fuel | | | Grand total |
|---|---|---|---|---|---|---|---|---|
| | Kerosene | LPG | Grid and Captive | Solar PV | Fuelwood | Charcoal | Others | |
| Unit | Ktoe | Ktoe | Ktoe | Ktoe | Ktoe | Ktoe | Ktoe | Ktoe |
| Space heating | – | – | 34.81 | – | – | 139.90 | – | 174.70 |
| Cooking and water heating | 856.64 | 113.33 | 687.09 | - | 77,853.79 | 2,109.02 | 2,700.96 | 84,320.83 |
| Lighting | 1,047.01 | – | 1,309.54 | 0.01 | – | – | – | 2,356.56 |
| Appliances | – | – | 1,262.96 | 0.26 | – | – | – | 1,263.22 |
| Total | 2,016.98 | | 3,294.67 | | 82,803.67 | | | 88,115.31 |

sector. The main drivers of the electricity demand growth are population increase, urbanization, economic growth, and improved access to electricity. The electricity supply mix is expected to change significantly in both scenarios, with more renewable sources and natural gas replacing oil and coal (Dioha and Emodi 2019).

## 3.3  Methodology

Modeling energy systems and how they relate to other aspects of a country, such as social and technical factors, is challenging. This is because predicting the future of any country and its sectors is not easy for modelers anywhere in the world. There are many uncertainties in how different factors within and outside the country affect the model choices and assumptions. To study how a changing system affects the outcomes, modelers often use scenarios that describe possible and reasonable futures. These scenarios do not give a fixed answer, but a range of possible outcomes. Scenarios can help modelers explore different options, compare different outcomes, and identify key uncertainties and risks. They can also help policymakers make informed decisions based on the best available knowledge.

This study used a bottom-up (End-Use) modeling approach, which considers all the sectors, sub-sectors, end-uses, and devices that use energy. This approach helps to understand the reasons for energy use in an economy better than the top-down (Econometric) modeling approach. The bottom-up approach is also useful because it does not depend on market behavior and production frontiers, which are often uncertain in developing countries (Urban et al. 2007). Moreover, the bottom-up approach can capture the effects of structural changes and technology-based policies, such as energy efficiency. The bottom-up approach can provide more detailed and accurate results for energy demand and supply analysis. It can also help to identify the potential for energy savings and emissions reductions in different scenarios.

The bottom-up modeling approach can show the details of the energy system technologies and processes. It can also compare different options for how these technologies and processes can change in the future (Pfenninger et al. 2014). Some examples of models that use the bottom-up approach are EnergyPLAN, unit commitment models, and HOMER (Lund 2006; Padhy 2004; Lambert et al. 2006). The bottom-up approach can also measure the environmental and social impacts of different energy system scenarios, such as how they affect greenhouse gas emissions, air quality, health, and energy access (Wiese et al. 2018). However, the bottom-up approach also has some challenges, such as finding and using reliable data, solving complex problems, and dealing with uncertainties (Pfenninger et al. 2014). The bottom-up approach can benefit from combining with other methods, such as top-down or hybrid approaches, to overcome some of these challenges. The bottom-up approach can also use different tools and techniques, such as optimization, simulation, or scenario analysis, to improve its results.

This study made use of the supply of LPG as a proxy for its demand. Preliminary results of a comparative analysis between the historical GDP growth rate and

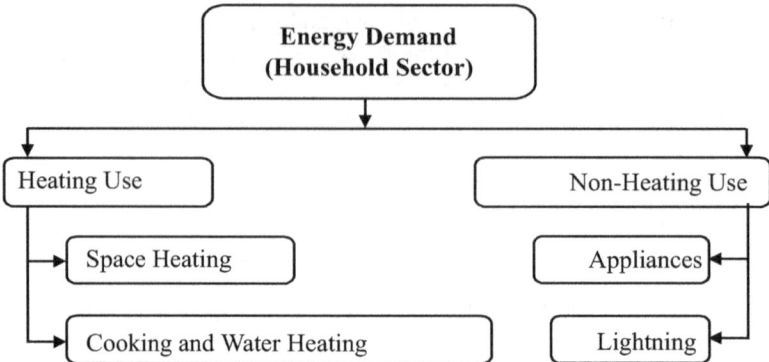

**Fig. 3.5** Nature of energy demand

historical growth rate in the supply of LPG in Nigeria show that the former had a reinforcing effect on the latter. Market-related conditions were not accounted for in the model since it was outside the scope of the study. Thus, the GDP growth rate was modeled to be the prime mover for the supply and consequently demand of LPG in Nigeria.

The dataset that was used for this study was obtained from the Energy Commission of Nigeria (ECN). This dataset was compiled by the ECN from several sources the National Bureau of Statistics, Central Bank of Nigeria, and the Department of Petroleum Resources.

### 3.3.1  Model Overview

The energy demand in the Nigerian household sector depends on two main factors: the end-use and the source of energy. The end-use can be either thermal or non-thermal, as shown in Fig. 3.5. The source of energy can be either fossil substitutable (such as kerosene and LPG), electricity, or traditional fuel (such as fuelwood, charcoal, and others). The amount of energy consumed by each household, the type and efficiency of the technology used, and other factors also affect the total energy demand in this sector. This study models the energy consumption in different forms of energy for different end-uses.

### 3.3.2  Energy Demand Computation Procedure in the Household Sector

The following equations show how the final demand in the household sector was calculated, based on the LEAP tool;

$$E \cdot D_{HH} = E \cdot D_{Space\,heating} + E \cdot D_{Cooking\,\&\,Water\,heating}$$
$$+ E \cdot D_{Lighting} + E \cdot D_{Appliances}$$

$$E \cdot D_{Space\,heating} = \sum_{Fuel} \{N * E \cdot I\}_{Fuel}$$

$$E \cdot D_{Cooking\,and\,Water\,heating} = \sum_{Fuel} \{N * E \cdot I\}_{Fuel}$$

$$E \cdot D_{Lighting} = \sum_{Fuel} \{N * E \cdot I\}_{Fuel}$$

$$E \cdot D_{Appliances} = \sum_{Type} N * E \cdot I_{type}$$

where;

$E \cdot D_{HH}$ = Energy Demand in the household sector expressed in Ktoe.

$E \cdot D_{Space\,heating}$ = Energy Demand for space heating expressed in Ktoe.

$E \cdot D_{Cooking\,and\,Water\,heating}$ = Energy demand for cooking and water heating expressed in Ktoe.

$E \cdot D_{Lighting}$ = Energy demand for lighting expressed in Ktoe.

$E \cdot D_{Appliances}$ = Energy demand for household appliances expressed in Ktoe.

$N$ = Number of households.

$E \cdot I_{Fuel}$ = Energy intensity of specific fuel type expressed in Ktoe/household.

$E \cdot I_{type}$ = Energy intensity of specific type of appliance expressed in Ktoe/household.

### 3.3.3 Development of Scenarios

This study examines how the population growth and the economic fluctuations of Nigeria affect the primary energy demand in different scenarios. The study considers four (4) main scenarios that depend on the average annual growth rates (AAGR) of the economy (at 2010 constant basic price), as shown in Table 3.3. These scenarios reflect different levels of economic development and energy consumption patterns. The study also explores a fifth scenario that focuses on the substitution of LPG for other fuels in cooking and water heating, based on one of the main scenarios. This scenario aims to assess the potential benefits of LPG adoption for households and the environment. The scenarios are;

i. **Business-as-usual Scenario**: In this scenario, the energy profile of the household sector is modeled. The Nigerian nominal GDP grows marginally at an annual average of 5% from 2021–2050.

ii. **Low Growth Scenario**: This scenario models the energy profile of the household sector given the GDP average annual growth rate of 6% from 2021–2050.

**Table 3.3** The average annual growth rate of the economy in each scenario

| Scenario | AAGR (%) |
|---|---|
| Business-as-usual | 5 |
| Low growth | 6 |
| Medium growth | 7 |
| High growth | 10 |
| LPG substitution policy scenario | 7 |

iii.  **Medium Growth Scenario**: Here the energy profile is modeled in the case that the average annual growth rate of the GDP over the period from 2021–2050 is 7%.
iv.  **High Growth Scenario**: The energy profile in this scenario is modeled when the average annual GDP growth rate is 10% between 2021–2050.
v.  **LPG Substitution Policy Scenario**: A fifth scenario called the LPG substitution policy scenario was also included; this scenario is based on the medium growth scenario. This was done because prevailing economic conditions in the medium growth scenario could aid in addressing the bottlenecks to the mainstreaming of LPG into the Nigerian household sector.

All scenarios have taken into consideration previous in-country interactions between the sectoral GDP of such sectors like the Industry sector (Manufacturing, Agriculture, Construction, and Mining sectors), the Transport sector, and the Service (Commercial) sector and the effects that such interactions have had on previous demand in the household sector. Also, across all the scenarios the population growth rate was kept constant at an annual average of 3.2%.

## 3.4   Results and Discussions

### 3.4.1   Energy Demand Scenarios in the Nigerian Household Sector

Table 3.4 shows the final energy demand for each scenario. The demand in this sector is projected to grow by different factors for each scenario by 2050. The BAU scenario has the highest growth factor of 2.38, followed by the LG scenario with 2.17, the MG scenario with 2.07, and the HG scenario with 1.84. This means that the more the economy grows (especially, the GDP/per capita as shown below), the more likely it is that households will use more modern fuels for cooking and water heating (as shown in Fig. 3.6).

**Table 3.4** Energy demand projections for the household sector

| Scenario (AAGR of GDP/capita) | 2022 | 2025 | 2030 | 2040 | 2050 | Ratio [2050/2010] |
|---|---|---|---|---|---|---|
| Unit | Ktoe | Ktoe | Ktoe | Ktoe | Ktoe | |
| BAU (1.33%) | 24,447.85 | 133,605.76 | 148,868.94 | 179,395.30 | 209,921.67 | 2.38 |
| LG (2.77%) | 23,813.95 | 131,070.19 | 143,163.91 | 167,351.35 | 191,538.79 | 2.17 |
| MG (4.01%) | 23,494.69 | 129,793.12 | 140,290.51 | 161,285.28 | 182,280.06 | 2.07 |
| HG (6.82%) | 22,814.51 | 127,072.42 | 134,168.94 | 148,361.98 | 162,555.02 | 1.84 |
| LPG (4.01) | 120,688.01 | 118,566.41 | 115,030.40 | 83,642.34 | 94,543.35 | 1.07 |

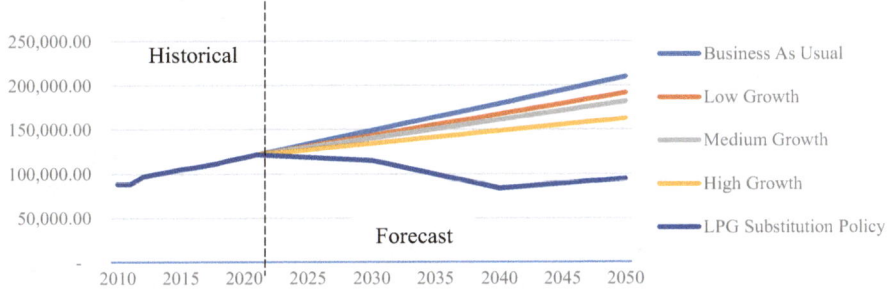

**Fig. 3.6** Historical and forecast demand for energy in the household sector

## 3.4.2 Business as Usual Scenario

The results from this scenario as presented in Table 3.4 and Fig. 3.7 show the household energy demand to have increased by a factor of 2.38 from 2010 to become 209,921.67 Ktoe by 2050. In this scenario, the AAGR of the economy is 4.60% per annum. With the steady growth profile of the population, the AAGR of the GDP/capita is expected to stay marginal at around 1.33%. Since income plays a role in transition within the household sector, this marginal growth rate in GDP/capita will not be particularly favorable to consuming more modern fuels.

Fuelwood is modeled in the scenario to continue to dominate in the energy mix with around 67% of households still using fuelwood by 2050. Kerosene and LPG are modeled to follow closely depending on the prevailing fuel price at around 8.93% and 17.68% respectively by 2050. Other fuels are modeled to remain marginal with a maximum of 2.1% of households using fuels from other sources by 2050.

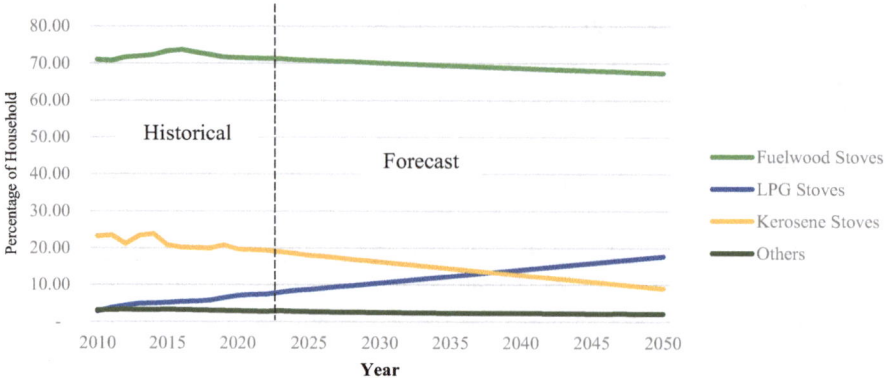

**Fig. 3.7** Dynamics of fuel used for cooking and water heating (2010–2050) in the BAU scenario

## 3.4.3  Low Growth Scenario

In the Low Growth scenario, demand for energy in the household sector as presented in Table 3.4 and Fig. 3.8 is shown to have increased by a factor of 2.17 from 2010 to become 191,538.79 Ktoe by 205,030. The LG scenario has more favorable economic conditions for the diffusion of modern fuels as compared to the BAU scenario. Here, the AAGR of the economy is modeled to be 6.08% per annum while the AAGR of the GDP/capita is 2.77%.

Although fuelwood is expected to continue to be the dominant fuel in the household sector, around 58.5% of households will use fuelwood in the LG scenario by 2050 in this scenario. LPG and Kerosene follow with approximately 21.4% and 9.4% of households respectively using these fuels by 2030. Other fuels stay marginal with a maximum of 1.7% of households using fuels from other sources by 2030.

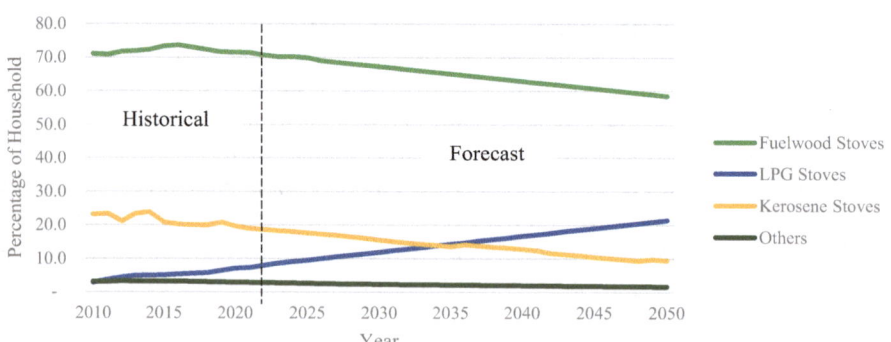

**Fig. 3.8** Dynamics of fuel used for cooking and water heating (2010–2050)

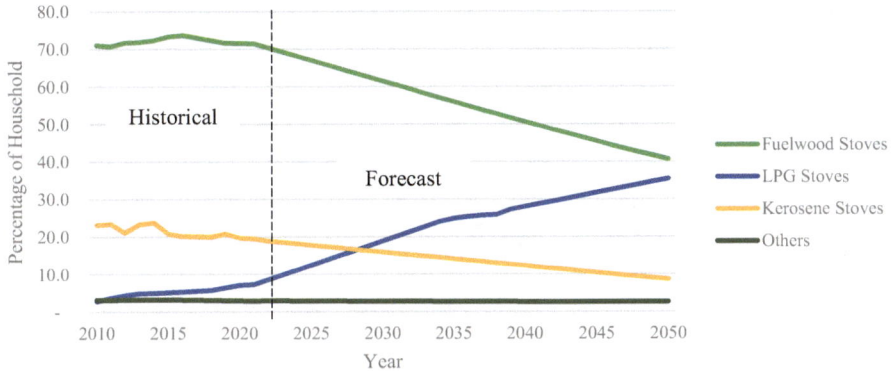

**Fig. 3.9** Dynamics of cooking and water heating (2010–2050) in Nigeria

### 3.4.4  Medium Growth Scenario

In this scenario, energy demand in the household sector is presented in Table 3.4 and Fig. 3.9. It is shown to have increased by a factor of 2.07 from 2010 to reach 182,280.06 Ktoe by 2030. The AAGR of the GDP, 7.37% per annum favors the mainstreaming of modern fuels when compared with the LG scenario. Also, the GDP/per capita is modeled to grow at an average of 4.01% per annum.

Fuelwood will still take the giant share of the energy mix with around 40.6% of households still using fuelwood by 2050. LPG and Kerosene follow with approximately 35.40% and 8.6% of households using these fuels respectively by 2050. While the other fuels remain marginal at around 2.7%. The slight increase recorded in the demand from other fuels is modeled to come from an increase in the use of electric cookstoves in the medium growth scenario.

### 3.4.5  High Growth Scenario

In this scenario, the economy grows very fast, at an average rate of 10.2% every year. The income per person also increases by about 7% every year on average. If Nigeria can keep up these good economic conditions, the households will need more energy. By 2050, the household energy demand will be 1.84 times higher than in 2010, reaching 162,555.02 Ktoe. Table 3.4 shows the numbers for this scenario. Figure 3.10 shows the percentage of households that use different fuels for cooking and water heating in this scenario.

This scenario assumes that the economy grows very fast. In this scenario, by 2050, about one-third of the households will still use wood as fuel. Most of the other households will use LPG or kerosene, with 50.7% and 10% respectively. The rest of

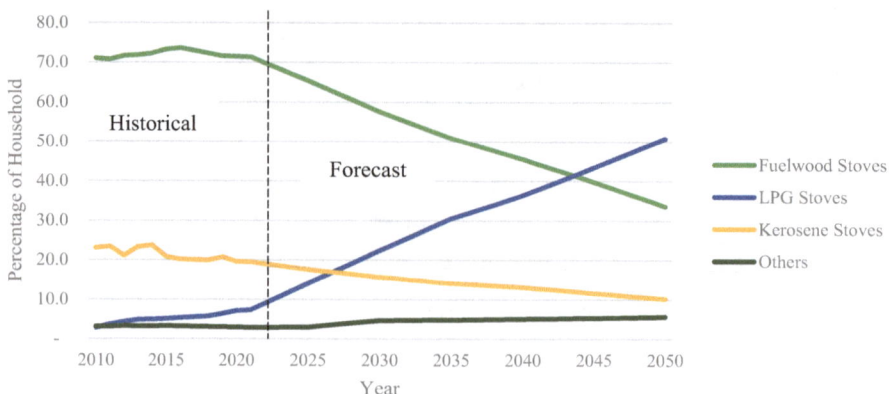

**Fig. 3.10**  Dynamics of fuel used for cooking and heating (2010–2050) in high-growth scenario

the households will use other fuels, such as electricity or solar, with only 5.7%. This is similar to the medium growth scenario.

### 3.4.6  LPG Substitution Policy Scenario in the Household Sector

LPG is a cleaner fuel than wood, and more Nigerian households are using it for cooking and water heating. Using LPG can help reduce energy poverty, climate change, and health problems. However, some barriers prevent LPG from becoming more popular in Nigerian households. These barriers need to be addressed by policymakers. Some of these barriers are:

*Barriers to the Adoption of LPG as the Fuel of Choice*

I.  **Cost:** LPG is mainly used by richer people in poor and lower-middle-income countries, according to a World Bank study of 10 countries (Kojima 2011). Nigeria is richer than these countries, so some of the poorer people might also use LPG. However, the World Bank study also found that LPG users need more than US$4200 per year at the current LPG prices, which is more than twice the average annual household income in Nigeria in 2017. Therefore, LPG is too expensive for most Nigerian households. Some of them still use wood as fuel, even though they can afford LPG. They could switch to LPG if the market conditions are better. They do not need financial help to do that.

II.  **Education:** How does education affect the choice of LPG as a fuel in households? People with higher education are more likely to choose LPG as their fuel. It also says that educating women about the costs and benefits of different fuels can help them switch to cleaner fuels like LPG.

*Market factors' role in the adoption of lpg as the fuel of choice*

I. **Inadequate Supply and Distribution Infrastructure:** The Nigerian Liquefied Natural Gas increased its LPG supply from 150,000 MT to 350,000 MT in five years, which made the LPG sector more active. Two new coastal depots with 12,500 MT capacity were built in Lagos and are being expanded by 8000 MT. More depots are being built or finished in Calabar, Port Harcourt, and Lagos. The NNPC/PPMC is also bringing back its inland depots to join the growing LPG market. The number of LPG refilling plants went up from 300 in 2009 to about 600 in 2016. The Department for Petroleum Resources (DPR) also allowed over 60 new LPG filling plants since 2012 (Department of Petroleum Resources [DPR] 2016). This will make LPG distribution easier in the country. However, most of the LPG facilities and use are in the south, where 65% of them are. This creates a regional imbalance because the north has more people who use traditional biomass for cooking and heating. They also have fewer biomass resources because of desertification. This surge in supply is laudable but it creates some form of regional imbalance since 65% of all LPG facilities and consumption are in the southern regions of the country. Ironically, the population who predominantly use traditional biomass for cooking and heating purposes and who have fewer biomass resources due to desertification are in the northern regions of the country.

II. **Use of Insufficient LPG Cylinders:** There are not enough cylinders in the country and most of them are old and unsafe (DPR 2016). Making cylinders in Nigeria is difficult because of bad business conditions in the past. Importing cylinders is also expensive because of the high taxes of 35%, which makes the cylinders too costly for the customers. More action on the part of policymakers, vendors, and users is needed to support the production of LPG cylinders in Nigeria.

III. **Safety:** There are safety concerns about using LPG as a cooking fuel in households. Some people are afraid of LPG because they have heard of cases where LPG exploded and caused fire. It is suggested that educating the consumers on how to prevent and handle leakages and fires from LPG would help them feel more confident. Also, having rules and inspections to check the quality of the cylinders would help avoid explosions caused by faulty cylinders.

To increase the use of LPG as a cooking fuel in Nigeria, the government of Nigeria launched a plan in November 2016 that aims to switch 4 million households to LPG by 2018, 10 million households by 2021, and 21 million households in the long term. This would increase the demand for LPG to 1.6 million MT by 2021 and 2.9 million MT by 2026 (DPR 2016). Figure 3.11 shows how the energy demand in the household sector would change if the plan is successful. It shows the percentage of households that use different fuels for cooking and water heating.

The LPG substitution policy scenario assumes a significant increase in the use of LPG as a fuel for cooking and water heating in households. By 2050, it is projected that about 41.7% of households, or 157 million people, will choose LPG for these purposes, compared to only 3.1% of households, or 5 million people, in 2010.

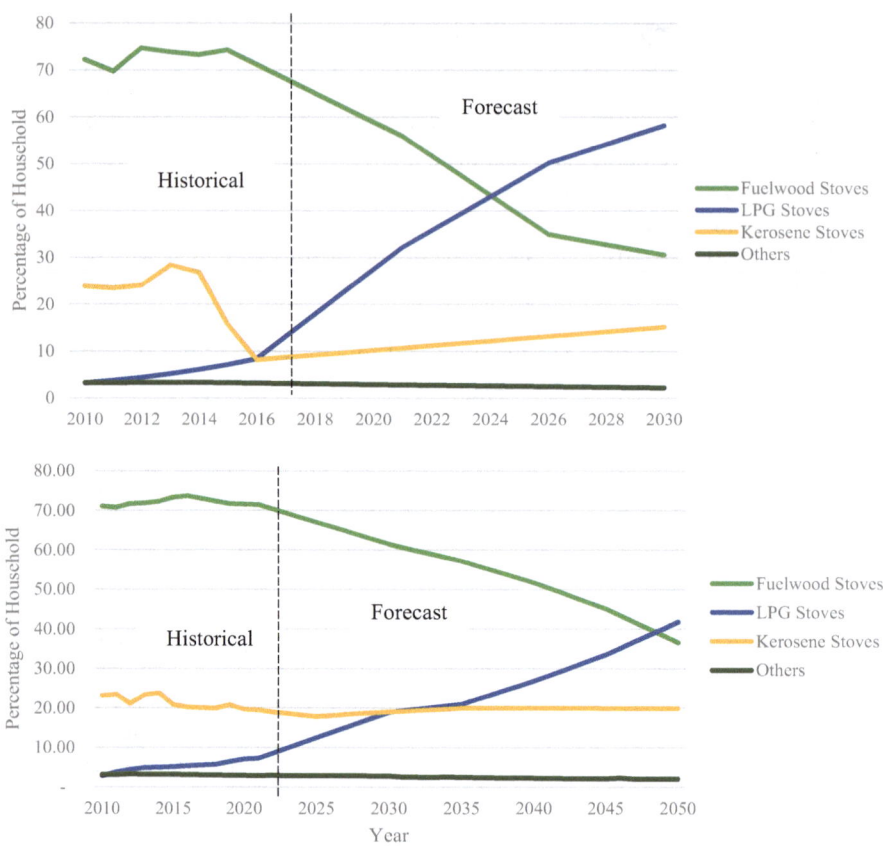

**Fig. 3.11** Dynamics of Projected use of fuel for cooking and water heating

## 3.5   Conclusion and Policy Implication

Energy transition has rightly been described in this study as the gradual rather than sudden change that unfolds over time. This change brings about more diversity to the energy system, as different sources, technologies, and actors are involved in the production, distribution, and consumption of energy. Energy transition in the twenty-first century has been focused on the need to make a switch from depleting fossil resources and exploring ways to harness energy from renewable sources, especially for the generation of electricity. This has led policymakers to emphasize solely utilizing RE for electricity generation, such as solar, wind, hydro, and biomass. However, this approach may neglect the environmental and social impacts of some RE sources, such as deforestation, land use change, and displacement of communities. Turning a blind eye to the fast depleting forest reserves could prove disastrous shortly given the global challenge of climate change. Therefore, energy transition should also consider the sustainability and equity of different energy options, and involve the participation and empowerment of local communities in decision-making processes. Our analysis elucidates the lacuna that currently exists in the energy profile of the Nigerian household sector and the urgent need to start to push for viable alternatives to the use of solid biomass for cooking in the household sector. Key insights for Nigerian energy policymakers are;

i. When it comes to energy policy design targeting the use of LPG, our analysis suggests that decision makers should target the Northern regions of the country for the construction of the distribution infrastructure of LPG.

ii. Regarding the incurred costs inherent in fuel switching, decision makers can focus policies on reducing the cost of equipment like gas burners and cylinders. Focusing subsidies on this equipment makes the subsidy less regressive over time since it is a one-off payment.

iii. Our analysis also strongly suggests that policies targeted at educating women on the costs and benefits of fuel switching would go a long way in the mainstreaming of LPG. This is because women are the ones who directly suffer the adverse health effects of the use of fuelwood in the kitchen.

iv. Our analysis likewise suggests that promoting the in-country production of LPG cylinders will lower the costs of cylinders. Lowering the cost makes it affordable to consumers. Policymakers can look into policy instruments targeted at attracting investors who may be looking to invest in the local production of cylinders.

v. Finally, the use of LPG is not without the inherent risks associated with its use. Decision makers should target policies at both the consumers and distributors to lower risks of explosion. Consumers should be educated on ways to prevent leakages and what can be done in the case of a leakage or explosion. The distributors should be made to ensure that cylinders still in circulation are safe for use.

Overall, consistent with The United Nations Sustainable Development Goal No. 7 (i.e., ensure universal access to clean and affordable electricity to all) (Dioha and Kumar 2020a, b; McCollum et al. 2018), this study contributes with a scenario-based analysis and empirical evidence to the effectiveness of LPG-based transition for the household sector of Nigeria that policymakers can benefit from.

Although the focus of this study was Nigerian household use of energy, findings of this study about the transition of energy from firewood to LPG are equally applicable to several countries of Africa and elsewhere where firewood is still the dominant source of energy for cooking in the household.

**Acknowledgements** Author would like to thank and acknowledge the work and support on the earlier draft of this chapter by Sunkanmi Dairo who collected the data financed by PAUWES, University of Tlemcen, Algeria, during his thesis project.

# References

Adams S, Acheampon A (2019) Reducing carbon emissions: the role of renewable energy and democracy. J Clean Prod 240:118245

AGECC (2010) Energy for a sustainable future: Report and recommendations. New York: United Nations

AGECC (2016) Energy for a sustainable future: summary report and recommendations. Retrieved from https://www.un.org/en/development/desa/policy/wess/wess_archive/2011wess.pdf

Ashagidigbi WM, Oyedepo SO, Kilanko OO (2020a) Energy access: a key factor for sustainable development goals in Nigeria. Energy Rep 6(2):1195–1208

Ashagidigbi WM, Babatunde BA, Ogunniyi AI, Olagunju KO, Omotayo AO (2020b) Estimation and determinants of multidimensional energy poverty among households in Nigeria. Sustainability 12:7332

Bekun FV, Alola AA, Gyamfi BA, Ampomah AB (2021) The environmental aspects of conventional and clean energy policy in sub-Saharan Africa: is the N-shaped hypothesis valid? Environ Sci Pollut Res 28(51):66695–66708. https://doi.org/10.1007/s11356-021-14758-w

Bouckaert S, Kim T, McNamara K (2019) Africa energy outlook 2019. World Economic Council and the International Energy Agency

Department of Petroleum Resources (2016) 2016 Oil and Gas Annual Report. Department of Petroleum Resources, Federal Republic of Nigeria. https://dpr.gov.ng/wp-content/uploads/2018/04/2016-Oil-Gas-Industry-Annual-Report.pdf

Dioha MO, Kumar A (2020) Achieving universal electricity access in Nigeria: a model-based analysis of challenges and opportunities. Energy Policy 137(111114)

Dioha MO, Emodi NV (2019) Investigating the impacts of energy access scenarios in the Nigerian household sector by 2030. Resources 8(3):127. https://doi.org/10.3390/resources8030127

Dioha M, Kumar A (2020b) Exploring sustainable energy transitions in sub-Saharan Africa residential sector: the case of Nigeria. Renew Sustain Energy Rev 17:109510

Engelman R (2009) In: Population, climate change, and women's lives. Worldwatch Institute

Faiella I, Lavecchia L, Miniaci R, Valbonesi P (2022) Household energy poverty and the "Just Transition". In: Zimmermann KF (ed) Handbook of labor, human resources and population economics, Springer, pp 1–20. https://doi.org/10.1007/978-3-319-57365-6_334-1

Grossman GM (2012) Population growth, sustainability, and endogenous technological change [Conference paper]. In: International conference on sustainable development

Hauff J, Kojima S, Trimble C (2019) Energy in Sub-Saharan Africa today: challenges and opportunities. In: Kojima S, Trimble C (eds) The state of electricity access report 2019. World Bank Group, pp 1–18

Hauff J, Bode A, Neumann D, Haslauer F (2014) Global energy transitions: a comparative analysis of key countries and implications for the international energy debate. World Energy Council and ATKearny. http://www.wec-france.org/DocumentsPDF/donnees/Global-Energy-Transitions-2014.pdf

https://africacheck.org/wp-content/uploads/2019/12/Africa_Energy_Outlook_2019.pdf

IRENA (2021) The renewable energy transition in Africa. Retrieved from https://www.irena.org/publications/2021/March/The-Renewable-Energy-Transition-in-Africa

Knight KW, Rosa EA (2011a) The environmental efficiency of well-being: a cross-national analysis. Soc Sci Res 40(3):931–949. https://doi.org/10.1016/j.ssresearch.2010.11.002

Knight K, Rosa E (2011b) Household dynamics and fuel consumption in developing countries: a cross-national analysis. Popul Environ 33(4):365–378

Kojima M (2011) The role of liquefied petroleum gas in reducing energy poverty. Extractive Industries for Development Series No. 25, World Bank

Kowsari R, Zerriffi H (2011) Three dimensional energy profile : a conceptual framework for assessing household energy use. Energy Policy 39(12):7505–7517

Lambe F, Jürisoo M, Wanjiru H, Senyagwa J (2015) Bringing clean, safe, affordable cooking energy to households across Africa : an agenda for action. Prepared by the Stockholm Environment

Institute, Stockholm and Nairobi, for the New Climate Economy. https://mediamanager.sei.org/documents/publications/nce-sei-2015-transforming-household-energy-sub-saharan-africa.pdf

Lambert T, Gilman P, Lilienthal P (2006) Micropower system modeling with homer. In: Farret FA, Simões MG (eds) Integration of alternative sources of energy. Wiley, pp 379–418

Leach G (1992) The energy transition. Energy Policy 20(2):116–123. https://doi.org/10.1016/0301-4215(92)90105-B

Li F, Liu X, Wang Y, Zhang Y (2020a) Renewable energy consumption, economic growth and environmental pollution in sub-Saharan Africa: evidence from panel data analysis. Energy Rep 6(2):1368–1378

Li X, Edwards D, Hosseini R, Costin G (2020b) A review on renewable energy transition in Australia: an updated depiction. J Clean Prod 242:118475

Lund H (2006) Large-scale integration of optimal combinations of PV, wind, and wave power into the electricity supply. Renew Energy 31(4):503–515. https://doi.org/10.1016/j.renene.2005.04.008

Maji I (2019) Impact of clean energy and inclusive development on $CO_2$ emissions in sub-Saharan Africa. J Clean Prod 240:118186

Mandelli S, Barbieri J, Mereu R, Colombo E (2016) Off-grid systems for rural electrification in developing countries: definition, classification, and a comprehensive literature review. Ren Sustain Energy Rev 58:1621–1646

McCollum DL, Echeverri LG, Busch S, Pachauri S, Parkinson S, Rogelj J et al (2018) Connecting the sustainable development goals by their energy interlinkages. Environ Res Lett 13:033006. https://doi.org/10.1088/1748-9326/aaafe3

Morgan G (Ed) (2012) Energy pathways for sustainable development. In: Johansson TB, Patwardhan A, Nakicenovic N, Gomez-Echeverri L (eds) Global energy assessment: toward a sustainable future. Cambridge University Press, pp 1205–1305

Müller T, Poganietz WR, Fahl U, Voß A (2021) The role of renewable energies in North Africa's low-carbon energy transition: a systematic literature review. Renew Sustain Energy Rev 145(111006)

Munasinghe M (2002) Analyzing the nexus of sustainable development and climate change: an overview. Working party on global and structural policies and the working party on development cooperation and environment. https://www.oecd.org/env/cc/2510070.pdf

National Bureau of Statistics (2011a) Annual abstract of statistics, 2011. Federal Republic of Nigeria, National Bureau of Statistics

National Bureau of Statistics (2012) Annual abstract of statistics, 2012. Federal Republic of Nigeria, National Bureau of Statistics

National Bureau of Statistics (2014a) Nigerian gross domestic product report quarter four. Federal Republic of Nigeria, National Bureau of Statistics

National Bureau of Statistics (2016a) Nigerian gross domestic product report quarter one 2016. Federal Republic of Nigeria, National Bureau of Statistics

National Bureau of Statistics (2017d) Nigerian gross domestic product report. Federal Republic of Nigeria, National Bureau of Statistics

National Bureau of Statistics (2011) Gross domestic product for Nigeria. National Bureau of Statistics, Federal Republic of Nigeria. https://nigerianstat.gov.ng/download/33

National Bureau of Statistics (2014) Power generation statistics 2010–2014:energy produced, turnover, and cost of operation. National Bureau of Statistics, the Federal Republic of Nigeria and Nigerian Electricity Regulatory Commission, Federal Republic of Nigeria

National Bureau of Statistics (2016) Power generation statistics 2015–Q2 2016: daily energy produced and sent out. National Bureau of Statistics, the Federal Republic of Nigeria and Nigerian Electricity Regulatory Commission, Federal Republic of Nigeria

National Bureau of Statistics (2017) Annual Abstract of Statistics, Vol. 1. National Bureau of Statistics, Federal Republic of Nigeria. https://www.nigerianstat.gov.ng/pdfuploads/ANNUAL%20ABSTRACT%20STATISTICS%20VOLUME-1.pdf

National Bureau of Statistics (2017) Annual abstract of statistics, vol 2. National Bureau of Statistics, Federal Republic of Nigeria. https://www.nigerianstat.gov.ng/pdfuploads/ANNUAL%20ABSTRACT%20VOLUME-2.pdf

National Bureau of Statistics (2017) Daily Energy Generated and Sent Out, (January) National Bureau of Statistics, the Federal Republic of Nigeria and Nigerian Electricity Regulatory Commission, Federal Republic of Nigeria. http://www.nigerianstat.gov.ng/download/502

Nsafon BEK, Same NN, Yakub AO, Chaulagain D, Kumar NM, Huh JS (2023) The justice and policy implications of clean energy transition in Africa. Front Environ Sci 11(1089391)

OECD (2019) Achieving clean energy access in sub-Saharan Africa. OECD Environment Policy Papers, No. 17. OECD Publishing. https://doi.org/10.1787/2257522f-en

Ogwumike FO, Abiona GA (2014c) Household energy use and determinants : evidence from Nigeria. Int J Energy Econ Policy 4(2):248–262

Ogwumike FO, Abiona G (2014) Household energy use pattern in Nigeria: evidence from cross river state [Conference paper]. In: International conference on advances in social science, economics and management study

Ogwumike FO, Abiona G (2014) Household energy use pattern in Nigeria: evidence from cross river state [Conference paper]. In: International conference on advances in social science, economics and management study. https://www.researchgate.net/publication/269713449_Household_Energy_Consumption_Pattern_and_SocioCultural_Dimensions_associated_with_it_A_Case_Study_of_Rural_India

Oxfam (2017) The energy challenge in sub-Saharan Africa: a guide for advocates and policy-makers: Part 1: generating energy for sustainable and equitable development. Oxfam International. https://www-cdn.oxfam.org/s3fs-public/file_attachments/bp-energy-challenge-africa-pt1-290317-en.pdf

Oyedepo SO, Adaramola MS, Paul SS, Oyedepo OJ (2012) Energy and sustainable development in Nigeria: the way forward. Energy Sustain Soc 2(1):15

Pachauri S, Spreng D (2011) Measuring and monitoring energy poverty. Energy Policy 39(12):7497–7504

Padhy NP (2004) Unit commitment—a bibliographical survey. IEEE transactions on systems, man, and cybernetics, Part C (Applications and Reviews). vol 34(2). pp 215–223. https://doi.org/10.1109/TSMCC.2003.818557

Pfenninger S, Hawkes A, Keirstead J (2014) Energy systems modeling for twenty-first-century energy challenges. Renew Sustain Energy Rev 33:74–86. https://doi.org/10.1016/j.rser.2014.02.003

Rapu CS, Adenuga AO, Kanya WJ, Abeng MO, Golit PD, Hilili MJ, Uba IA, Ochu ER (2015) Analysis of energy market conditions in Nigeria. Central Bank of Nigeria [CBN]. https://www.cbn.gov.ng/out/2017/rsd/analysis of energy.pdf

REN21 (2016) UN report on renewable energy policy network for the 21st century. https://www.ctc-n.org/network/network-members/renewable-energy-policy-network-21st-century. Accessed on 09 Nov 2019

Schlag N, Zuzarte F (2008) Market barriers to clean cooking fuels in sub-Saharan Africa: A review of literature. Stockholm: SEI

Sepp S, Ahlborg H, Arvidson A, Eckerberg K (2014) Household energy consumption in Sub-Saharan Africa: the role of gender equality and environmental sustainability in achieving the millennium development goals [Conference paper]. In: World renewable energy congress XIII and exhibition 2014—innovation in Europe renewable energy research and technology. https://www.researchgate.net/publication/266507731_Household_Energy_Consumption_in_SubSaharan_Africa_The_Role_of_Gender_Equality_and_Environmental_Sustainability_in_Achieving_the_Millennium_Development_Goals

Sepp S, Sepp C, Mundhenk M (2014) Towards sustainable modern wood energy development. Federal Ministry for Economic Cooperation and Development and Global Bioenergy Partnership. http://www.globalbioenergy.org/fileadmin/user_upload/gbep/docs/giz2015-en-report-wood-energy.pdf

Su B, Heshmati A, Geng Y, Yu X (2019a) A review of global energy transition: evolutionary stages and driving forces. Renew Sustain Energy Rev 100(1):296–318

Su M, Yue W, Liu Y, Tan Y, Shen Y (2019b) Integrated evaluation of urban energy supply security: a network perspective. J Clean Prod 209:461–471

Sunil K, Govinda RT (2014) Household energy consumption pattern and socio-cultural dimensions associated with it: a case study of rural India [Conference paper]. In: International conference on advances in energy research 2013

Sunil K, Govinda RT (2014) Household energy consumption pattern and socio-cultural dimensions associated with it: a case study of rural India [Conference paper]. In: International conference on advances in energy research 2013. https://www.researchgate.net/publication/269713449_Household_Energy_Consumption_Pattern_and_SocioCultural_Dimensions_associated_with_it_A_Case_Study_of_Rural_India

Sunil M, Govinda T (2014) Household cooking fuel choice and adoption of improved cookstoves in developing countries (WPS6903). World Bank

Toole S (2015) From the 'Energy Ladder' to 'Fuel Stacking' [Blog post]. Energypedia. https://energypedia.info/wiki/From_the_%27Energy_Ladder%27_to_%27Fuel_Stacking%27

Tsan M, Gupta G, Wendle J, Collett P, Jamal F, Kanchwala Y, Jayannavar P, Whalley N, Intrator K, Wade I, Wade K, Kay E, Juneja N, Ankit M (2014) Clean and improved cooking in Sub-Saharan Africa. World Bank, Africa Renewable Energy Access Program and Energy Sector Management Assistance Program

TWI (2021) What is energy transition? (Definition, Benefits, and Challenges). https://www.twi-global.com/technical-knowledge/faqs/energy-transition

Urban F, Benders RMJ, Moll HC (2007) Modeling energy systems for developing countries. Energy Policy 35(6):3473–3482

van der Mandelli S, Barbieri J, Mereu R, Colombo E (2016b) Off-grid systems for rural electrification in developing countries: definitions, classification, and a comprehensive literature review. Renew Sustain Energy Rev 58(1):1621–1646

Victoria-Foye AO, Oluwatosin AO, Oluwaseunfunmi OS (2020) Impacts of population, climate change and governance on economic growth in Nigeria [Conference paper]. In: International conference on sustainable development

Wiese F, Hilpert S, Kaldemeyer C, Pleßmann G (2018) A qualitative evaluation approach for energy system modeling frameworks. Energy Sustain Soc 8(1), Article number: 13. https://doi.org/10.1186/s13705-018-0154-3

World Bank (2021) Population growth (annual %)—Nigeria | Data

World Economic Forum (2020) Energy transition 101: getting back to basics for transitioning to a low-carbon economy. https://www3.weforum.org/docs/WEF_Energy_Transition_101_2020.pdf

World Economic Forum (2021) 5 key lessons for energy transition from COVID-19 recovery. https://www.weforum.org/agenda/2021/04/5-key-lessons-for-energy-transition-from-covid-19-recovery/

Worlddata.info (n.d.) Energy consumption in Nigeria. Retrieved from https://www.worlddata.info/africa/nigeria/energy-consumption.php

WRI (2021) Energy access is key to sub-saharan africa's economic recovery [Blog post]. World Resources Institute. https://www.wri.org/insights/energy-access-key-sub-saharan-africas-economic-recovery

Yergin D, Iniesa JE, Samuelsen SG (2013) The global energy transition: a roadmap to 2050. https://www.energy-transitions.org/sites/default/files/ETC_Roadmap_to_2050.pd

Yergin D, Gross S, Meyer N, Tilemann-Dick L (2013) Energy vision 2013: energy transition : past and future (Issue January). International Energy Agency. http://www3.weforum.org/docs/WEF_EN_EnergyVision_Report_2013.pdf

# Chapter 4
# Bamboo Gasification for Sustainable Energy and Rural Development in Uganda

**Abstract** Bamboo gasification is a promising social innovation for sustainable energy and rural development in Uganda. Bamboo is a fast-growing and versatile plant that can provide a renewable and reliable source of biomass for gasification. Bamboo gasification can generate electricity and heat, as well as other valuable products, such as biochar, bio-oil, or bio-fertilizer. Bamboo gasification can also contribute to various Sustainable Development Goals, such as poverty reduction, climate change mitigation, forest conservation, and job creation. However, bamboo gasification faces many challenges and barriers in Uganda, such as a lack of awareness, research, investment, and support. This chapter aims to explore the potential and challenges of bamboo gasification in Uganda and to propose a conceptual and methodological framework for its promotion and commercialization. The chapter reviews the literature and case studies on biomass gasification, bamboo cultivation, and utilization, technical and environmental performance of gasification technologies, applications and markets for syngas and its derivatives, socio-economic and policy aspects of bamboo gasification, and sensitivity analysis of feasibility assessment. Based on the review, the chapter presents a conceptual model that depicts the main components and relationships of the bamboo gasification system and its environment. The chapter also proposes a 5-step approach that guides the decision makers to apply the conceptual model to any region or country that has bamboo resources. The chapter concludes with some limitations of the conceptual model and some directions for future research.

***What is in for the readers of this chapter?*** In this chapter, readers can find a conceptual and methodological framework for studying bamboo gasification as a social innovation for sustainable energy and rural development in Uganda and other countries. Researchers can use the conceptual model and the 5-step approach to guide their research questions, hypotheses, data collection, and analysis methods. Policymakers can use the findings and recommendations of the feasibility assessment to support and facilitate the promotion and commercialization of bamboo gasification projects. Practitioners can learn from the literature and case studies on bamboo cultivation and utilization, gasification technologies, syngas applications and markets,

© The Author(s), under exclusive license to Springer Nature Switzerland AG 2024     63
H. Qudrat-Ullah, *Exploring the Dynamics of Renewable Energy and Sustainable Development in Africa*, Advances in African Economic, Social and Political Development, https://doi.org/10.1007/978-3-031-48528-2_4

socio-economic and policy aspects, and sensitivity analysis. University's research libraries and research centers can benefit from the knowledge and insights on bamboo gasification and its potential and challenges.

## 4.1 Introduction

Energy access is a key driver for sustainable development and poverty alleviation, as it enables economic activities, social services, and environmental protection (IEA 2020a; World Bank 2021). However, according to the International Energy Agency (IEA), about 759 million people still lack access to electricity and 2.6 billion people rely on traditional biomass for cooking in 2020 (IEA 2020b). Most of these people live in rural areas of sub-Saharan Africa and Asia, where grid extension is often not feasible or cost-effective (Borowski et al. 2021; Menin et al. 2021a, b). Therefore, decentralized renewable energy solutions, such as biomass gasification, can play a vital role in providing clean, reliable, and affordable energy services to these populations (Basu et al. 2020a, b; Menin et al. 2021a, b).

Biomass gasification is a thermochemical process that converts solid biomass into a combustible gas mixture (syngas) that can be used for various applications, such as electricity generation, heating, cooling, or fuel production. Biomass gasification has several advantages over the direct combustion of biomass, such as higher efficiency, lower emissions, and greater flexibility (Hrbek et al. 2021). Moreover, biomass gasification can utilize a wide range of feedstocks, including agricultural residues, forest wastes, energy crops, and municipal solid waste (Jafri et al. 2020). Biomass gasification is considered a promising technology for achieving the UN Sustainable Development Goal 7 (SDG7) of ensuring affordable, reliable, sustainable, and modern energy for all by 2030 (Waldheim et al. 2022). The global biomass gasification market is expected to grow from $90.56 billion in 2021 to $98.63 billion in 2022 at a compound annual growth rate (CAGR) of 8.9% and reach $139.49 billion in 2026 at a CAGR of 9.1% (GlobeNewswire 2022).

However, biomass gasification also faces some challenges and opportunities that need to be addressed for its successful deployment and commercialization. Some of the key challenges include the high capital costs of gasification plants and equipment, the low quality and variability of biomass feedstocks, the formation and removal of tar and other contaminants from syngas, and the integration and optimization of gasification systems with downstream applications (Drift et al. 2015; U.S. Department of Energy 2023). Some of the opportunities include the development and adoption of emerging gasification technologies that offer higher efficiency, lower emissions, and greater flexibility than conventional gasifiers, the utilization of syngas for producing hydrogen and other valuable chemicals and fuels, and the implementation of supportive policies and incentives that promote biomass gasification as a sustainable energy solution (Sikarwar et al. 2016; Waldheim et al. 2022; Zafar et al. 2017).

Among the various types of biomass feedstocks, bamboo has emerged as a promising option for biomass gasification due to its fast growth rate, high productivity, low maintenance cost, and multiple environmental benefits. Bamboo is a perennial grass that can grow up to 30 m in height and can be harvested every 3–5 years without replanting. Bamboo can sequester carbon dioxide from the atmosphere and store it in its biomass and soil. Bamboo can also prevent soil erosion, conserve water resources, and enhance biodiversity. Furthermore, bamboo can provide multiple socio-economic benefits to rural communities, such as income generation, employment creation, food security, and handicraft production (Antonini et al. 2016).

Bamboo gasification is a process that converts bamboo biomass into syngas, a mixture of hydrogen, carbon monoxide, methane, and other gases that can be used for various applications, such as electricity generation, heating, cooling, or fuel production. Bamboo gasification has several advantages over direct combustion of bamboo, such as higher efficiency, lower emissions, and greater flexibility. Bamboo gasification can also utilize different types of bamboo feedstocks, such as bamboo waste, bamboo chips, or bamboo pellets (Powermax Gasifiers 2023).

Bamboo gasification is considered a sustainable and renewable energy technology that can contribute to the mitigation of climate change and the achievement of the UN Sustainable Development Goals (SDGs). Bamboo gasification can reduce greenhouse gas emissions by displacing fossil fuels and by capturing biogenic carbon dioxide from the syngas. Bamboo gasification can also improve energy access and security for rural areas that lack reliable electricity supply or grid connection. Bamboo gasification can also create value-added products from bamboo biomass, such as biochar, bio-oil, or biofuels (Sikarwar et al. 2016; Truong and Le 2014).

However, bamboo gasification also faces some challenges and opportunities that need to be addressed for its successful deployment and commercialization. Some of the key challenges include the high capital costs of gasification plants and equipment, the low quality and variability of bamboo feedstocks, the formation and removal of tar and other contaminants from syngas, and the integration and optimization of gasification systems with downstream applications (Drift et al. 2015; U.S. Department of Energy 2023). Some of the opportunities include the development and adoption of emerging gasification technologies that offer higher efficiency, lower emissions, and greater flexibility than conventional gasifiers, the utilization of syngas for producing hydrogen and other valuable chemicals and fuels, and the implementation of supportive policies and incentives that promote bamboo gasification as a sustainable energy solution (Hrbek et al. 2021; Sikarwar et al. 2016; Zafar et al. 2017).

Bamboo is a versatile plant that can offer a sustainable solution for biomass gasification and rural development in Uganda. However, despite its abundant availability and numerous benefits, bamboo has not been exploited as an energy source in Uganda. Uganda is a landlocked country in East Africa with a population of about 45 million people and an electrification rate of about 28% (Antonini et al. 2016). Uganda has a rich biomass resource base, estimated at 500 PJ per year, but most of it is used inefficiently for traditional cooking and heating purposes, resulting in high greenhouse gas emissions and deforestation (Zafar et al. 2017). Uganda also has a large potential for bamboo cultivation and utilization, as it has suitable climatic and soil conditions and existing bamboo plantations and nurseries. However, bamboo is mainly used for low-value applications such as furniture making, basket weaving, fencing, and handicrafts (Truong and Le 2014).

Bamboo gasification is a process that converts bamboo biomass into syngas, a mixture of hydrogen, carbon monoxide, methane, and other gases that can be used for various applications, such as electricity generation, heating, cooling, or fuel production. Bamboo gasification has several advantages over direct combustion of bamboo, such as higher efficiency, lower emissions, and greater flexibility. Bamboo gasification can also utilize different types of bamboo feedstocks, such as bamboo waste, bamboo chips, or bamboo pellets (Powermax Gasifiers 2023).

Bamboo gasification is considered a sustainable and renewable energy technology that can contribute to the mitigation of climate change and the achievement of the UN Sustainable Development Goals (SDGs). Bamboo gasification can reduce greenhouse gas emissions by displacing fossil fuels and by capturing biogenic carbon dioxide from the syngas. Bamboo gasification can also improve energy access and security for rural areas that lack reliable electricity supply or grid connection. Bamboo gasification can also create value-added products from bamboo biomass, such as biochar, bio-oil, or biofuels (Sikarwar et al. 2016; Truong and Le 2014).

To promote bamboo gasification as an energy source in Uganda, some challenges and opportunities need to be addressed. Some of the key challenges include the high capital costs of gasification plants and equipment, the low quality and variability of bamboo feedstocks, the formation and removal of tar and other contaminants from syngas, and the integration and optimization of gasification systems with downstream applications (Drift et al. 2015; U.S. Department of Energy 2023). Some of the opportunities include the development and adoption of emerging gasification technologies that offer higher efficiency, lower emissions, and greater flexibility than conventional gasifiers (Hrbek et al. 2021), the utilization of syngas for producing hydrogen and other valuable chemicals and fuels (Sikarwar et al. 2016), the implementation of supportive policies and incentives that promote bamboo gasification as a sustainable energy solution (Zafar et al. 2017), and the involvement of local communities in bamboo cultivation and utilization for income generation, employment creation, food security, and handicraft production (Africanews 2022; Forest Machine Magazine 2020; Monitor 2022).

There is a lack of comprehensive and updated information and analysis on the potential and challenges of bamboo gasification in Uganda, as well as the opportunities and recommendations for its promotion and commercialization. Most of the existing studies and reports on biomass gasification in Uganda focus on other feedstocks, such as agricultural residues, forest wastes, or municipal solid waste, and do not consider the specific characteristics and benefits of bamboo. Moreover, most of the existing studies and reports on bamboo gasification are based on experiences from other countries and regions, such as China, India, or Southeast Asia, and do not reflect the local context and conditions of Uganda. Therefore, there is a need for a more in-depth and contextualized study on bamboo gasification in Uganda that can provide useful insights and guidance for policy makers, practitioners, researchers, and stakeholders.

Specifically, in this chapter, we assess the techno-economic feasibility of a Bamboo Biomass Gasifier Power Plant (BGPP) for rural electrification within an innovative sustainable rural development concept (Nantongo 2017). The concept involves the establishment of a bamboo plantation that supplies feedstock to the BGPP and also provides multiple co-benefits to the local community. The BGPP is based on an air-blown bubbling fluidized bed (BFB) gasifier coupled with a gas engine generator that produces electricity for off-grid rural areas. The BGPP also produces biochar as a by-product that can be used as a soil amendment or a fuel. We use a cost–benefit analysis and a sensitivity analysis based on data from previous BGPP installations in Asia and Africa to evaluate the economic viability of the concept. We also estimate the environmental and social impacts of the concept by using indicators such as greenhouse gas emissions reduction, carbon sequestration potential, income generation potential, and Sustainable Development Goals (SDGs) achievement.

The chapter is organized as follows: Sect. 4.2 provides an overview of the bamboo biomass gasification technology and its applications and background information and energy dynamics of Uganda. Section 4.3 describes the methodology used for the techno-economic analysis of the BGPP concept. Section 4.4 presents the results and discussion of the analysis. Section 4.5 concludes the chapter with some policy recommendations and future research directions.

## 4.2 Overview of Bamboo Gasification Technology and Sustainable Development

### 4.2.1 Bamboo and Sustainable Development

Social innovation is the process of developing and implementing novel solutions to social problems that are more effective, efficient, sustainable, or just than existing solutions (Phills et al. 2008). Social innovation can address various challenges and opportunities in different sectors and contexts, such as health, education, environment, energy, and governance (Mulgan et al. 2007). Bamboo-based development is a concept that uses bamboo as a renewable and versatile resource for economic growth, environmental protection, and social welfare (Lobovikov et al. 2017a, b). Bamboo-based development can be considered a form of social innovation, as it can provide multiple benefits for various stakeholders, such as income generation, employment creation, food security, handicraft production, carbon sequestration, soil conservation, water management, and biodiversity enhancement (Mwampamba et al. 2021a, b). Africa has great potential for social innovation through bamboo-based development, as it has abundant natural bamboo resources, rich cultural diversity, and immense challenges and opportunities. However, Africa also faces some barriers and constraints that hinder the full exploitation of bamboo resources and the realization of bamboo-based development. These include the lack of awareness, knowledge, skills, technology, policy, and investment in the bamboo sector (Truong and Le 2014; Zafar et al. 2017).

Some possible ways to promote the adoption of social innovation through bamboo-based development in Africa are:

- Raising awareness and education on the benefits and uses of bamboo for various purposes, such as energy, construction, furniture, handicrafts, food, medicine, etc. (Mwampamba et al. 2019).
- Providing technical and financial support for bamboo cultivation, harvesting, processing, and marketing, as well as for developing innovative and sustainable bamboo products and services.
- Creating platforms and networks for collaboration and knowledge exchange among different stakeholders, such as farmers, entrepreneurs, researchers, policymakers, NGOs, etc., who are involved or interested in bamboo-based development.
- Encouraging participatory and inclusive approaches that involve the local communities and youth in identifying and solving their social problems using bamboo as a tool and a resource.
- Aligning bamboo-based development with the national and regional policies and strategies for renewable energy, climate change mitigation, poverty reduction, and sustainable development (United Nations Africa Renewal Magazine 2016; Mwampamba et al. 2021a, b; Lobovikov et al. 2017a).

In this section, we have discussed the concept and potential of social innovation through bamboo-based development in Africa. We have highlighted the multiple benefits and opportunities that bamboo can offer for addressing various social problems and enhancing economic, environmental, and social welfare in the continent. We have also identified some of the challenges and barriers that limit the exploitation and utilization of bamboo resources and the implementation of bamboo-based development initiatives. Finally, we have suggested some possible ways to promote the adoption of social innovation through bamboo-based development in Africa, by raising awareness, providing support, creating platforms, encouraging participation, and aligning policies. The next section will explore the specific case of bamboo gasification as a social innovation for sustainable energy and rural development in Africa.

### 4.2.2  Bamboo-Based Biomass Gasification Technology and Its Applications

Bamboo-based biomass gasification is a thermochemical process that converts bamboo biomass into a combustible gas mixture (syngas) that can be used for various applications, such as electricity generation, heating, cooling, or fuel production.

Bamboo-based biomass gasification has several advantages over direct combustion of bamboo, such as higher efficiency, lower emissions, and greater flexibility. Bamboo-based biomass gasification can also utilize different types of bamboo feedstocks, such as bamboo waste, bamboo chips, or bamboo pellets (Powermax Gasifiers 2023).

Bamboo-based biomass gasification is considered a sustainable and renewable energy technology that can contribute to the mitigation of climate change and the achievement of the UN Sustainable Development Goals (SDGs). Bamboo-based biomass gasification can reduce greenhouse gas emissions by displacing fossil fuels and by capturing biogenic carbon dioxide from the syngas. Bamboo-based biomass gasification can also improve energy access and security for rural areas that lack reliable electricity supply or grid connection. Bamboo-based biomass gasification can also create value-added products from bamboo biomass, such as biochar, bio-oil, or biofuels (Sikarwar et al. 2016; Truong and Le 2014).

However, bamboo-based biomass gasification also faces some challenges and opportunities that need to be addressed for its successful deployment and commercialization. Some of the key challenges include the high capital costs of gasification plants and equipment, the low quality and variability of bamboo feedstocks, the formation and removal of tar and other contaminants from syngas, and the integration and optimization of gasification systems with downstream applications (Drift et al. 2015; U.S. Department of Energy 2023). Some of the opportunities include the development and adoption of emerging gasification technologies that offer higher efficiency, lower emissions, and greater flexibility than conventional gasifiers (Hrbek et al. 2021), the utilization of syngas for producing hydrogen and other valuable chemicals and fuels (Sikarwar et al. 2016), the implementation of supportive policies and incentives that promote bamboo-based biomass gasification as a sustainable energy solution (Zafar et al. 2017), and the involvement of local communities in bamboo cultivation and utilization for income generation, employment creation, food security, and handicraft production (Africanews 2022; Forest Machine Magazine 2020; Monitor 2022).

The literature review reveals that bamboo-based biomass gasification is a promising technology with multiple benefits and applications. However, there is still a lack of comprehensive framework and analysis of several issues regarding, for example, (i) the assessment and optimization of the quality and quantity of bamboo feedstocks in different regions and contexts, and the development of suitable pretreatment and storage methods to improve their gasification performance, (ii) the comparison and evaluation of the efficiency, emissions, and costs of different gasification technologies and configurations for bamboo feedstocks, and the identification of the optimal operating parameters and conditions for each technology and application, and (iii) The integration and optimization of gasification systems with downstream applications, such as power generation, heating, cooling, or fuel production, and the development of innovative and sustainable products and services from syngas and its derivatives Therefore, future research ab investigate theses issues to promote the adoption of bamboo-based biomass gasification as a social innovation for sustainable energy and rural development.

### 4.2.3  Bamboo Gasification for Sustainable Energy and Rural Development in Africa

Bamboo gasification has several advantages over the direct combustion of biomass, such as higher efficiency, lower emissions, and greater flexibility (Bridgwater 2012). Moreover, bamboo gasification can utilize a wide range of bamboo species and forms, including culms, branches, leaves, and residues (Basu et al. 2020a, b). Bamboo gasification can also produce biochar as a by-product, which can be used as a soil amendment or a fuel (Menin et al. 2021a, b).

Bamboo gasification has been widely practiced in Asia, especially in China and India, where it has been used for rural electrification and industrial applications (Basu et al. 2020a, b). However, bamboo gasification is still relatively new and underdeveloped in Africa, where bamboo resources are abundant but underutilized (Partey et al. 2017a, b). According to the International Network for Bamboo and Rattan (INBAR), an intergovernmental organization that promotes the development of bamboo and rattan for economic and environmental gains, Africa has about 3.6 million hectares of natural bamboo forests and 1.2 million hectares of planted bamboo, mainly in Ethiopia, Madagascar, Tanzania, and Uganda (INBAR 2018a, b).

Bamboo gasification has the potential to contribute to sustainable energy and rural development in Africa by providing clean, reliable, and affordable energy services to off-grid populations, enhancing rural economic growth by creating income opportunities from bamboo production and processing, reducing greenhouse gas emissions by substituting fossil fuels and traditional biomass with renewable energy sources, and improving soil fertility and carbon sequestration by applying biochar to degraded lands (Basu et al. 2020a, b; Menin et al. 2021a, b; Partey et al. 2017a, b). However, there are also some challenges and barriers that hinder the widespread adoption and scaling-up of bamboo gasification in Africa. These include:

- Lack of awareness and knowledge about bamboo resources and their potential uses among policymakers, investors, farmers, and consumers (Partey et al. 2017a, b).
- Lack of supportive policies and incentives that promote bamboo cultivation, management, harvesting, transportation, and utilization (Partey et al. 2017a, b).
- Lack of technical capacity and skills for bamboo gasification technology development, operation, maintenance, and quality control (Basu et al. 2020a, b).
- Lack of access to finance and markets for bamboo products and services (Partey et al. 2017a, b).
- Lack of adequate infrastructure and logistics for bamboo supply chain development (Basu et al. 2020a, b).
- Lack of comprehensive data and information on bamboo resources availability, distribution, characteristics, and performance (Partey et al. 2017a, b).

Therefore, there is a need for more research and development on bamboo gasification for sustainable energy and rural development in Africa. Some of the potential research areas that could be intensified include:

- Assessment of bamboo resources availability, distribution, characteristics, and performance in different agroecological zones of Africa (Partey et al. 2017a, b).
- Development of appropriate bamboo gasification technologies that are suitable for the local conditions and needs of African countries (Basu et al. 2020a, b).
- Evaluation of the techno-economic feasibility and environmental impacts of bamboo gasification systems for different applications and scales (Menin et al. 2021a, b).
- Analysis of the social acceptability and institutional arrangements for bamboo gasification systems implementation and operation (Partey et al. 2017a, b).
- Identification of best practices and lessons learned from successful bamboo gasification projects in Asia and other regions (Basu et al. 2020a, b).

Bamboo gasification is a promising technology that can provide sustainable energy and rural development in Africa, where bamboo resources are abundant but underutilized. Bamboo gasification can offer multiple benefits, such as providing clean, reliable, and affordable energy services, enhancing rural economic growth, reducing greenhouse gas emissions, and improving soil fertility and carbon sequestration. However, there are also some challenges and barriers that hinder the widespread adoption and scaling-up of bamboo gasification in Africa, such as a lack of awareness, knowledge, policies, incentives, technical capacity, skills, finance, markets, infrastructure, and data. Bamboo gasification can be a strong pillar of Africa's future green economy if supported by appropriate policies and actions from governments, the private sector, civil society, and international organizations. Future research could focus on topics such as:

- How to optimize the design, operation, and maintenance of bamboo gasification systems for different climatic and socio-economic conditions in Africa?
- How to assess the life cycle impacts and benefits of bamboo gasification systems on greenhouse gas emissions, air quality, water resources, biodiversity, and human health in Africa?
- How to enhance the value chain and market development of bamboo products and services in Africa, including bamboo gasification systems, biochar, electricity, heat, cooling, and fuel?
- How to improve the governance and institutional arrangements for bamboo gasification systems implementation and operation in Africa, including stakeholder engagement, capacity building, financing mechanisms, and regulatory frameworks?
- How to integrate bamboo gasification systems with other renewable energy sources and technologies, such as solar, wind, hydro, biogas, and microgrids, to increase the resilience and reliability of energy supply in rural areas of Africa

## 4.3  Background Information and Energy Dynamics in Uganda

### 4.3.1  An Overview of Background Information, Population, and Economy of Uganda

Uganda is a country in East-Central Africa that does not have a coastline. It covers an area of 241,551 square kilometers (93,263 square miles) and borders Kenya to the east, South Sudan to the north, the Democratic Republic of the Congo to the west, Rwanda to the southwest, and Tanzania to the south. The equator crosses Uganda, which lies between 4°N and 2°S latitude and 29° and 35°E longitude. Uganda is part of the Great Lakes region of Africa and has several lakes within its territory, including Lake Victoria (the largest lake in Africa), Lake Albert, and Lake Edward. Most of Uganda is within the Nile basin, with Lake Victoria being the main source of the White Nile. About 18% of Uganda's land area is water (World Bank 2020a, b, c).

Uganda has a warm tropical climate that varies according to altitude and location. The climate is influenced by the Inter-Tropical Convergence Zone and different wind patterns that bring moist or dry air from the Atlantic Ocean, the Indian Ocean, or the neighboring countries. Uganda has two rainy seasons (March to May and September to November) and two dry seasons (December to February and June to August). The average annual rainfall ranges from 900 to 1600 mm, with more rain in the south and west than in the north and east. The average annual temperature is about 23 °C, with a maximum of 29 °C and a minimum of 17 °C. The hottest months are December to February, while the coolest months are June to August. The northeast of Uganda has a semi-arid climate with less rain and higher temperatures than the rest of the country. The northwest of Uganda, near Lake Albert and the Albert Nile River, has very high temperatures (up to 40 °C) and low rainfall in the dry season. The central south of Uganda, around the equator, has a tropical savannah climate with stable temperatures (27 °C to 29 °C) throughout the year (WorldAtlas 2020; Wikipedia 2021).

Uganda has a population of about 47.1 million people (World Bank 2020a, b, c). The population is ethnically and culturally diverse, with more than 40 languages spoken. The official languages are English and Swahili, but Luganda, Runyoro, Luo, Lusoga, and Runyankole are also widely used. The majority of the population is rural and depends on agriculture for their livelihood. Uganda has a high population growth rate of 3.2% per year, which poses challenges for social services, environmental management, and economic development (World Bank 2020a, b, c).

Uganda's economy is largely based on agriculture, which accounts for about 22% of the gross domestic product (GDP) and employs about 70% of the labor force (World Bank 2020a, b, c). The main crops are coffee, tea, bananas, maize, cassava, beans, and cotton. Uganda also has abundant natural resources, such as gold, oil, and minerals. The oil sector is expected to boost the economy shortly, as commercial production is projected to start in 2025 (World Bank 2020a, b, c). Uganda also has a vibrant service sector, which contributes about 50% of the GDP and includes tourism, trade, finance, and telecommunications.

Uganda's economy has shown resilience and recovery in recent years, despite the shocks of the COVID-19 pandemic, locust invasion, floods, and political unrest. The GDP growth rate was 4.6% in 2022 and is estimated to reach 5.7% in 2023 (World Bank 2020a, b, c). The inflation rate was 2.2% in 2022 and is expected to remain low in 2023 (World Bank 2020a, b, c). The poverty rate (measured at $1.90 per day) declined from 32.9% in 2016 to 29.4% in 2019 (World Bank 2020a, b, c). However, the COVID-19 crisis reversed some of the poverty reduction gains and increased inequality and vulnerability among the population. The GDP per capita was $847 in 2020 and is projected to increase to $884 in 2021 (MacroTrends 2021).

Uganda's economy faces several challenges and opportunities for sustainable development. Some of the challenges include high public debt (51.8% of GDP in 2020), low revenue mobilization (12.1% of GDP in 2020), weak governance and institutional capacity, corruption and human rights violations, environmental degradation and climate change impacts, low human capital development, and service delivery, limited economic diversification and structural transformation, and regional instability and security threats (World Bank 2020a, b, c; Heritage Foundation 2023). Some of the opportunities include: harnessing the demographic dividend and youth potential, leveraging the oil sector and other natural resources for inclusive growth, investing in infrastructure and digital technology for connectivity and innovation, enhancing regional integration and trade for market access and competitiveness, promoting social protection and resilience for poverty reduction and social cohesion, and strengthening policy reforms and partnerships for governance and accountability (World Bank 2020a, b, c; Heritage Foundation 2023).

In conclusion, Uganda is a country with a large and diverse population and a growing and resilient economy. However, the country also faces many challenges and risks that could undermine its development prospects and goals. Uganda needs to address its structural and institutional constraints, enhance its human and natural capital, and seize its opportunities for social and economic transformation. The next section will provide an overview of the dynamics of energy and gasification in Uganda, which is a key sector for sustainable development and social innovation in the country.

### 4.3.2  Dynamics of Energy Supply and Demand in Uganda

Uganda's electricity sector is crucial for its economic growth, and the country needs to explore new energy sources to transform its rural areas, where over 80% of the population lives (Bucholz and Da Silva 2010). However, as of 2016, most of Uganda's energy supply came from biomass, and only a small fraction of the population had access to electricity, as shown in Fig. 5.4 below (IRENA 2020). While nearly 60% of the urban population had electricity access, only 18% of the rural population did, though this number was increasing (SEforALL Africa Hub 2023; World Bank 2023; Eberhardt et al. 2005).

The share of energy consumption by category in Uganda in 2022 can be estimated based on the latest available data from various sources. According to the International Renewable Energy Agency (IRENA 2020), the total energy supply (TES) in Uganda in 2019 was 966,391 terajoules (TJ), of which 92% came from renewable sources and 8% from non-renewable sources. The renewable energy supply was mainly composed of bioenergy (98%), followed by hydro/marine (2%), solar direct (<1%), and geothermal (<1%). The non-renewable energy supply was mainly composed of oil (100%), followed by coal and others (<1%), and gas (<1%). The total energy consumption (TFEC) in Uganda in 2019 was 708,000 TJ, of which 99% came from renewable sources and 1% from non-renewable sources. Renewable energy consumption was mainly composed of bioenergy (98%), followed by geothermal (2%), solar direct (<1%), and hydro/marine (<1%). The non-renewable energy consumption was mainly composed of oil (100%), followed by coal and others (<1%), and gas (<1%).

The consumption by sector in Uganda in 2019 was dominated by households (76%), followed by industry (13%), other sectors (11%), and transport (<1%). The household sector consumed mostly bioenergy (99%), followed by electricity (1%). The industry sector consumed mostly bioenergy (97%), followed by electricity (3%). The other sectors consumed mostly bioenergy (99%), followed by electricity (1%). The transport sector consumed mostly oil (100%).

Assuming that the trends of energy supply and consumption in Uganda remain similar to those observed in the previous years, the share of energy consumption by category in Uganda in 2022 can be projected as follows:

- Renewable sources: 99% of TFEC, mainly bioenergy.
- Non-renewable sources: 1% of TFEC, mainly oil.
- Household sector: 76% of TFEC, mainly bioenergy.
- Industry sector: 13% of TFEC, mainly bioenergy.
- Other sectors: 11% of TFEC, mainly bioenergy.
- Transport sector: < 1% of TFEC, mainly oil.

The electricity generation mix of Uganda as of 2022 can be estimated based on the latest available data from various sources. According to the International Energy Agency (IEA 2022), the total electricity generation in Uganda in 2020 was 4.8 terawatt-hours (TWh), of which 87% came from hydropower, 9% from solar, and 4% from oil. The IEA projects that the total electricity generation in Uganda will increase to 5.9 TWh in 2022, of which 84% will come from hydropower, 11% from solar, and 5% from oil. According to other sources, Uganda is expected to add more hydropower capacity in 2022 with the completion of the Karuma Hydro Power Dam, which will have a capacity of 600 megawatts (MW). The dam is expected to boost Uganda's total capacity by 44.7% and potentially leave over 1000 MW of excess power generation capacity (International Trade Administration 2022). Uganda is also expected to increase its solar capacity in 2022 with the commissioning of several mini-grid and off-grid projects supported by various donors and investors

(Off-Grid Solar Energy Market Assessment Brief—Uganda, 2023). Therefore, the electricity generation mix of Uganda as of 2022 can be projected as follows: (i) Hydropower: 84% of total generation, mainly from large-scale dams, (ii) Solar: 11% of total generation, mainly from mini-grid and off-grid systems, and (iii) Oil: 5% of total generation, mainly from diesel power plants.

The dynamics of energy supply and demand in Uganda show that the country has a high potential for renewable energy sources, especially bioenergy, and hydropower, but also faces challenges in expanding electricity access and diversifying its energy mix. While the government has set ambitious targets and policies to increase the use of modern renewable energy and improve energy efficiency, there is still a gap between the current situation and the desired outcomes. One of the technologies that could help bridge this gap is bamboo gasification, which can offer multiple benefits for sustainable energy and rural development in Uganda. Bamboo gasification can provide clean, reliable, and affordable energy services to remote communities, boost rural economic growth by creating income opportunities from bamboo production and processing, cut greenhouse gas emissions by replacing fossil fuels and traditional biomass with renewable energy sources, and enhance soil fertility and carbon sequestration by applying biochar to degraded lands. Therefore, there is a need for more research and development on bamboo gasification in Uganda, as well as more support from the government and the international community to promote its adoption and scaling up.

### 4.3.3  Biomass and Bioenergy Technology Development in Uganda

Uganda has a lot of land that is not good for growing food crops, but it can be used for growing woody crops that can be turned into energy (Bucholz and Da Silva 2010). However, Uganda has not adopted or developed bioenergy technology much. Most people use biomass in traditional ways, such as cooking for their families (Okello et al. 2013a, b). They do not know or understand how to use biomass in clean and modern ways (BMAU 2015). Also, they think that modern biomass systems are too expensive to use (Okello et al. 2013a, b).

The Ministry of Finance, Planning and Economic Development of Uganda said that biomass can be a big source of clean and modern energy/electricity in Uganda (BMAU 2015). They estimated that biomass residues from crops, animals, and forests could produce 258.29 PJ/year of energy (Okello et al. 2013a, b). They also said that Uganda has about 51 million tonnes of biomass that can be used for bioenergy production without harming the environment (BMAU 2015). This biomass comes from different sources, such as trees, bushes, papyrus, reeds, farm waste, agro-processing waste, and grass. Trees are the main source of biomass, with 26.3 million tonnes, followed by bushes, with 10 million tonnes (BMAU 2015). Uganda also promised to increase its forest cover from 15 to 18% by 2021 and to 24% by 2040 (CIF 2023).

Uganda's energy supply and consumption in 2019 were mostly based on renewable sources, especially bioenergy, according to IRENA (2020). The total energy supply (TES) was 966,391 TJ, with 92% from renewable sources and 8% from non-renewable sources. The main renewable source was bioenergy (98%), followed by hydro/marine (2%), solar direct (<1%), and geothermal (<1%). The main non-renewable source was oil (100%), followed by coal and others (<1%), and gas (<1%). The total energy consumption (TFEC) was 708,000 TJ, with 99% from renewable sources and 1% from non-renewable sources. The main renewable source was bioenergy (98%), followed by geothermal (2%), solar direct (<1%), and hydro/marine (<1%). The main non-renewable source was oil (100%), followed by coal and others (<1%), and gas (<1%). This shows that Uganda has a high potential for renewable energy development, especially in the areas of solar, hydro, and geothermal. However, it also faces some challenges, such as a low electrification rate, high dependence on biomass for cooking, and limited infrastructure and investment. Uganda needs to adopt policies and strategies that can promote the sustainable use of renewable energy and reduce the environmental and social impacts of non-renewable energy.

The consumption by sector in Uganda in 2019 was dominated by households (76%), followed by industry (13%), other sectors (11%), and transport (<1%). The household sector consumed mostly bioenergy (99%), followed by electricity (1%). The industry sector consumed mostly bioenergy (97%), followed by electricity (3%). The other sectors consumed mostly bioenergy (99%), followed by electricity (1%). The transport sector consumed mostly oil (100%).

The electricity generation mix of Uganda as of 2022 can be estimated based on the latest available data from various sources. According to the International Energy Agency (IEA, 2022), the total electricity generation in Uganda in 2020 was 4.8 TWh, of which 87% came from hydropower, 9% from solar, and 4% from oil. The IEA projects that the total electricity generation in Uganda will increase to 5.9 TWh in 2022, of which 84% will come from hydropower, 11% from solar, and 5% from oil.

As the biomass situation analysis indicates, Uganda's main challenge is not the insufficient supply of biomass but rather the technology to utilize the diverse forms of biomass. Uganda has the potential to have surplus biomass for energy purposes for the next three decades and possibly beyond. Biomass is poised to be a significant source of clean energy, such as electricity and biofuels if appropriate technologies are adopted. However, the level of biomass technology recognition and prioritization in the policy and institutional framework is low, as well as the awareness and adoption among the end-users. The high upfront investment costs and the lack of adequate financing mechanisms are also barriers to the development of biomass technology in Uganda. Hence, more research and development on biomass technology in Uganda is needed to promote its adoption and scaling up.

### 4.3.4 Bamboo Gasification and Sustainable Energy in Uganda

By converting bamboo biomass into a combustible gas, bamboo gasification can offer multiple benefits for sustainable energy and rural development in Uganda. It can supply clean, reliable, and affordable energy services to remote communities, boost rural economic growth by creating income opportunities from bamboo production and processing, cut greenhouse gas emissions by replacing fossil fuels with renewable energy sources, and enhance soil fertility and carbon sequestration by applying biochar to degraded lands (Basu et al. 2020a, b; Menin et al. 2021a, b; Partey et al. 2017a, b).

One of the initiatives that promote bamboo gasification and sustainable energy in Uganda is the Divine Bamboo Group Ltd (DBGL), an eco-friendly agroforestry and energy company that was founded in 2016 by Divine Nabaweesi. DBGL aims to stop deforestation in Uganda through bamboo forestry and use the plant as a base for clean biomass energy. DBGL has developed a briquette from bamboo waste that is 30% cheaper than the charcoal on the market. DBGL also engages smallholder farmers in bamboo cultivation and processing, providing them with income opportunities and environmental benefits. DBGL has set out expansion plans to increase its production capacity of briquettes from 240 tonnes currently to 2400 tonnes annually in ten years, as well as to establish 5000 acres of bamboo plantations and 1,000,000 seedlings production in the next five years (The Independent 2023).

The Earth Energy Company Limited (EECL) is a Ugandan firm that works on bamboo gasification and sustainable energy in Uganda. It got a grant of US $993,000 from the Sustainable Energy Fund for Africa (SEFA) in 2016 to start the first biomass gasification project in Uganda. The project will use agricultural residues to make a gas that can produce 20 Megawatts of power for the national grid. The project will also help 15,000 farmers earn more money by selling their agricultural residues to the project. The project will create 6,000 new jobs near Gulu Town in Uganda. The project will also protect the forests, reduce air pollution, and empower rural women. The project supports the Renewable Energy Policy for Uganda and the New Deal on Energy for Africa strategy (African Development Bank Group 2016).

The International Network for Bamboo and Rattan (INBAR) is an intergovernmental organization that supports bamboo gasification and sustainable energy in Uganda. Since 2002, INBAR has been working with various bamboo stakeholders in Uganda, providing technical assistance, capacity building, policy advocacy, and knowledge sharing on bamboo and rattan. INBAR has helped to create the Uganda National Bamboo Association (UNBA), a network that aims to promote bamboo as a strategic resource for sustainable development. INBAR has also facilitated the development of the National Bamboo Strategy and Action Plan for Uganda (2018–2022), which provides a roadmap for enhancing the contribution of bamboo to national development goals, such as poverty reduction, climate change mitigation and adaptation, biodiversity, and green industrialization (INBAR 2018a, b).

Another aspect of bamboo gasification and sustainable energy in Uganda is the potential for carbon finance and climate action. Bamboo is a fast-growing plant that can sequester large amounts of carbon dioxide from the atmosphere and store it in its biomass and soil. Bamboo also has a low carbon footprint compared to other biomass sources, as it requires minimal inputs and processing. Bamboo gasification can therefore reduce greenhouse gas emissions by displacing fossil fuels and traditional biomass with renewable energy sources. Moreover, bamboo gasification can generate carbon credits that can be sold in the voluntary or compliant carbon markets, providing an additional revenue stream for the project developers and investors. According to a study by INBAR and the World Agroforestry Centre (ICRAF), bamboo gasification projects in Africa could potentially generate up to 7.5 million tons of carbon dioxide equivalent (tCO2e) per year, equivalent to US $37.5 million per year at a carbon price of US $5 per tCO2e (INBAR & ICRAF, 2014).

In summary, Bamboo is a versatile and renewable plant that can offer multiple benefits for sustainable energy and rural development in Uganda. By converting bamboo biomass into a combustible gas, bamboo gasification can provide clean, reliable, and affordable energy services to remote communities, boost rural economic growth by creating income opportunities from bamboo production and processing, cut greenhouse gas emissions by replacing fossil fuels and traditional biomass with renewable energy sources, and enhance soil fertility and carbon sequestration by applying biochar to degraded lands (Basu et al. 2020a, b; Menin et al. 2021a, b; Partey et al. 2017a, b). Several initiatives have been launched to promote bamboo gasification and sustainable energy in Uganda, such as the Divine Bamboo Group Ltd (DBGL), the Earth Energy Company Limited (EECL), the International Network for Bamboo and Rattan (INBAR), and the potential for carbon finance and climate action. These initiatives demonstrate the feasibility, viability, and desirability of bamboo gasification as a technology that can contribute to Uganda's green economy and national development goals.

## 4.4  Building a Conceptual Model for Bamboo Gasification for Sustainable Energy and Rural Development in Uganda

Based on our literature analysis, we have developed a conceptual model for bamboo gasification for sustainable energy and rural development in Africa. This model is shown in Fig. 4.1. This model illustrates the relationships between the dependent variable (the level of adoption or diffusion of bamboo gasification as a social innovation) and the independent variables (the factors that influence the adoption or diffusion of bamboo gasification). The independent variables are grouped into four categories: individual, organizational, institutional, and environmental. The individual factors include the awareness, knowledge, skills, attitudes, and preferences of the potential adopters or users of bamboo gasification. The organizational factors include the availability, accessibility, affordability, and quality of bamboo gasification technology, products, and services. The institutional factors include the policies, regulations, incentives, and support systems that enable or constrain the development and implementation of bamboo gasification initiatives. The environmental factors include the availability and suitability of bamboo resources, the demand and market for bamboo gasification products and services, and the socio-economic and ecological impacts of bamboo gasification.

In Fig. 4.1, the arrow from the top-right box (*Bamboo gasification system characteristics*) to the bottom box (*Sustainable energy and rural development indicators*) is solid and has a plus sign (+), indicating a positive direct relationship. The arrow from the left-top box (*Bamboo resource characteristics*) to the top-right box is dashed and has a plus sign (+), indicating a positive moderating relationship. The arrow from the bottom-left box (*Bamboo gasification system barriers and enablers*) to the bottom-right box is dotted and has a plus sign (+), indicating a positive mediating relationship. The model also assumes that there are interactions and feedback loops among the independent variables that can amplify or moderate their effects on the dependent variable. The model can be used to guide the research questions, hypotheses, data collection, and analysis methods for studying bamboo gasification as a social innovation for sustainable energy and rural development in Uganda.

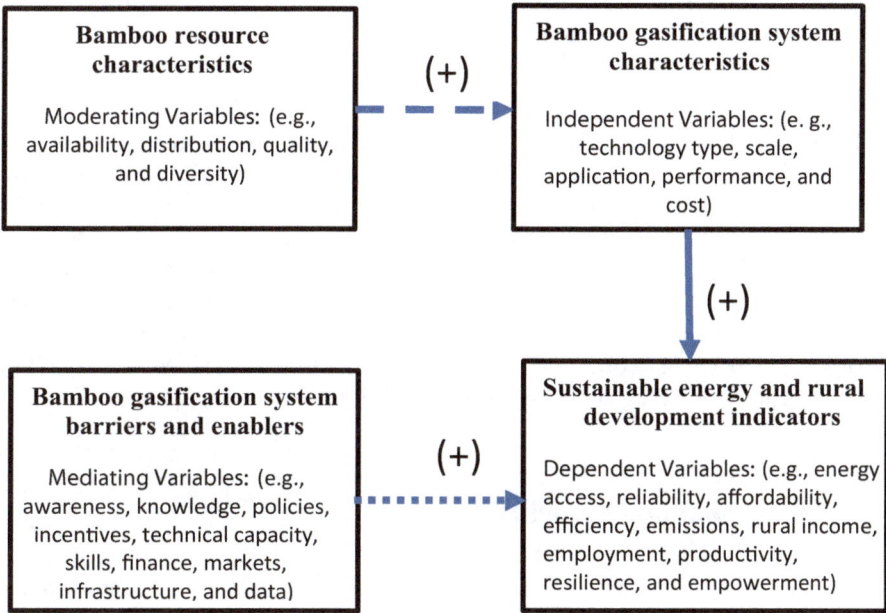

**Fig. 4.1** A conceptual model for bamboo gasification for sustainable energy and rural development in Uganda

### 4.4.1   How to Apply the Conceptual Model for Bamboo Gasification?

The conceptual model for bamboo gasification that we have developed is not only relevant for Uganda but also for other regions and countries that have abundant bamboo resources. However, how can the decision makers use this model to plan and implement their bamboo gasification projects? To answer this question, we propose a 5-step approach that can guide the decision makers to evaluate the socio-economic viability of their bamboo gasification projects. The 5-step approach consists of the following steps:

1.  Identify the potential sites for bamboo gasification based on the availability of bamboo resources, the demand for electricity and heat, and the environmental and social factors.
2.  Estimate the technical and economic performance of the bamboo gasification system, including the capital and operating costs, the energy output, and the greenhouse gas emissions.
3.  Assess the socio-economic benefits of the bamboo gasification project, such as income generation, job creation, poverty reduction, and the improvement of living standards.
4.  Compare the bamboo gasification project with alternative energy sources, such as diesel generators, grid electricity, or solar panels, in terms of cost-effectiveness, reliability, and sustainability.
5.  Conduct a sensitivity analysis to examine how the results of the feasibility assessment change under different scenarios and assumptions.

Figure 4.2 illustrates the 5-step approach for applying the conceptual model for bamboo gasification. The 5-step approach for applying the conceptual model for bamboo gasification is a useful tool for decision makers who want to explore the potential of bamboo as a sustainable energy source and a driver of rural development. The approach can help them to identify the most suitable sites, estimate the technical and economic performance, assess the socio-economic benefits, compare with alternative energy sources, and conduct a sensitivity analysis. The approach is flexible and adaptable to different contexts and conditions and can be applied to any region or country that has bamboo resources, be it in Africa or elsewhere.

## 4.5   Discussion and Conclusion

Sustainable energy and rural development are two interrelated topics that have gained increasing attention in recent years. Sustainable energy is the energy that can be used to meet the current needs without affecting the ability of future generations to meet their own needs. (Dincer and Rosen 2012). Sustainable energy sources are usually renewable, such as solar, wind, hydro, biomass, or geothermal. Sustainable energy

**Step-1 Estimation of load:** The electricity demand of the region/district is estimated based on the population, household size, electrification rate, and consumption patterns of different sectors and end-users.

**Step-2 Estimation of the required Bamboo quantity:** The biomass feedstock requirement of the gasification plant is estimated based on the plant capacity, efficiency, availability, and calorific value of bamboo.

**Step-3 Estimation of the required land for Bamboo plantation:** The land area needed to cultivate bamboo sustainably for the gasification plant is estimated based on the bamboo yield, density, rotation period, and harvesting rate.

**Step-4 Choice of the specifications of the installation:** The technical specifications of the gasification plant are selected based on the optimal size, type, configuration, and performance of the system.

**Step-5 Economic cost and investment analysis:** The economic viability of the gasification project is analyzed by estimating the capital cost, annual feedstock cost, annual operation and maintenance cost, levelized cost of electricity, net present value, internal rate of return, payback period, and discounted payback period. A systematic sensitivity analysis on the variables of interest to support decision makers is conducted.

**Fig. 4.2** Steps for the assessment of socio-economic feasibility of bamboo gasification projects

can also imply the efficient and rational use of energy, as well as the reduction of greenhouse gas emissions and other environmental impacts. Rural development refers to the improvement of the economic, social, and environmental conditions of rural areas, which often face challenges such as poverty, inequality, isolation, lack of infrastructure and services, and vulnerability to climate change. Rural development can also imply the empowerment and participation of rural people and communities in the management of their affairs, as well as the preservation and enhancement of their cultural and natural heritage.

Sustainable energy and rural development are closely linked, as they can mutually reinforce each other. On one hand, sustainable energy can contribute to rural development by providing reliable and affordable electricity and heat for households, businesses, industries, and public services. Sustainable energy can also create income and employment opportunities for rural people, especially through the cultivation and utilization of biomass resources, such as bamboo. Sustainable energy can also improve the living standards, health, education, gender equality, and social cohesion of rural people. Sustainable energy can also help to mitigate and adapt to climate change, by reducing greenhouse gas emissions and enhancing the resilience of rural areas to extreme weather events. On the other hand, rural development can contribute to sustainable energy by providing abundant and diverse sources of renewable energy, such as biomass, solar, wind, hydro, or geothermal. Rural development can also increase the demand and access to sustainable energy for rural people and communities, by improving their income levels, awareness levels, and infrastructure conditions. Rural development can also support the participation and collaboration of various stakeholders in the planning and implementation of sustainable energy projects.

In this chapter, we have presented a conceptual model for bamboo gasification as a means of sustainable energy and rural development in Uganda. We have also proposed a 5-step approach for applying the conceptual model to any region or country that has bamboo resources. We have argued that bamboo gasification can offer multiple benefits, such as increasing forest cover, reducing greenhouse gas emissions, creating jobs and income, improving living standards, and enhancing energy security and reliability. We have also discussed the challenges and opportunities of bamboo cultivation and utilization, as well as the technical and economic aspects of biomass gasification. We have shown that bamboo is a suitable feedstock for gasification, with high energy content and low emissions. We have also reviewed the existing literature and case studies on biomass gasification, especially in developing countries, and highlighted the potential of fixed bed gasification technology with catalytic gas cleaning.

Our conceptual model and 5-step approach ensure that the local population benefits from the candidate bamboo gasification project in several ways. First, they involve the participation and consultation of the local stakeholders, such as the farmers, the consumers, the local authorities, and the civil society organizations, in the planning and implementation of the project. This ensures that their needs, preferences, and concerns are taken into account and addressed. Second, they provide opportunities for the local population to benefit from the project directly or indirectly, such as by becoming bamboo growers, suppliers, workers, or entrepreneurs. This creates income and employment opportunities for the local people, as well as enhances their skills and capacities. Third, they assess the socio-economic impacts of the project on the local population, such as the effects on income, poverty, living standards, health, education, gender equality, and social cohesion. This helps to identify and mitigate any potential negative impacts or risks and to maximize the positive impacts or benefits. Fourth, they compare the project with alternative energy sources, such as diesel generators, grid electricity, or solar panels, in terms of cost-effectiveness, reliability, and sustainability. This helps to ensure that the project is affordable and accessible for the local population, as well as environmentally friendly and socially acceptable. Fifth, they conduct a sensitivity analysis to examine how the results of the feasibility assessment change under different scenarios and assumptions. This helps to account for uncertainties and variations in factors such as bamboo availability, energy demand, market prices, policy changes, or climate change. This also helps to identify and prepare for any potential challenges or opportunities that may arise in the future.

Some of the policy implications of our research are:

- The government should recognize and support bamboo gasification as a viable and beneficial option for sustainable energy and rural development in Uganda. The government should provide incentives, subsidies, tax exemptions, or grants for bamboo gasification projects, as well as facilitate access to land, water, credit, and technology.
- The government should establish and enforce clear and consistent regulations and standards for bamboo gasification, such as on the quality of feedstock, syngas, and by-products, emissions and waste management, safety and health measures, and monitoring and evaluation mechanisms.
- The government should promote and facilitate the participation and collaboration of various stakeholders, such as the farmers, the consumers, the local authorities, the civil society organizations, the private sector, the research institutions, and the international partners, in the planning and implementation of bamboo gasification projects. The government should ensure that the interests and concerns of the local population are taken into account and addressed.
- The government should invest in and support research and development on bamboo gasification, such as on the improvement of bamboo varieties, cultivation practices, gasification technologies, syngas applications, and socio-economic impacts. The government should also disseminate and share the research findings and best practices with the relevant stakeholders.

Our conceptual model, given in Fig. 4.1, guides the research questions, hypotheses, data collection, and analysis methods for studying bamboo gasification as a social innovation for sustainable energy and rural development in Uganda in the following way:

- The conceptual model depicts the main components and relationships of the bamboo gasification system, such as the bamboo cultivation, the gasification plant, the energy output, the socio-economic benefits, and the environmental impacts. It also shows the external factors that influence the system, such as the policy, market, technology, and climate contexts. The conceptual model provides a holistic and systemic view of the bamboo gasification system and its interactions with its environment.
- The research questions are derived from the conceptual model by identifying the key aspects and issues that need to be investigated and understood. For example, some of the research questions are: How feasible is bamboo gasification in Uganda? What is the technical and economic performance of the bamboo gasification system? What are the socio-economic benefits and environmental impacts of the bamboo gasification system? How do the policy, market, technology, and climate context affect the bamboo gasification system? How can the bamboo gasification system be implemented and scaled up in Uganda?

- The hypotheses are formulated from the conceptual model by making assumptions or predictions about the expected outcomes or relationships of the bamboo gasification system. For example, some of the hypotheses are: Bamboo gasification is a viable and beneficial option for sustainable energy and rural development in Uganda. Bamboo gasification can increase forest cover, reduce greenhouse gas emissions, create jobs and income, improve living standards, and enhance energy security and reliability. Bamboo gasification is influenced by various factors such as bamboo availability, energy demand, market prices, policy changes, or climate change.
- The data collection methods are designed from the conceptual model by selecting the appropriate sources and types of data that can answer the research questions and test the hypotheses. For example, some of the data collection methods are literature reviews, case studies, surveys, interviews, focus groups, observations, experiments, simulations, or secondary data analysis.
- The analysis methods are chosen from the conceptual model by applying the suitable techniques and tools that can process and interpret the data collected. For example, some of the analysis methods are descriptive statistics, inferential statistics, regression analysis, cost–benefit analysis, impact assessment, scenario analysis, or sensitivity analysis.

Some limitations of our conceptual model are:

- It may not capture all the possible components and relationships of the bamboo gasification system, or all the external factors that influence it. There may be some aspects or issues that are overlooked or oversimplified in the conceptual model.
- It may not reflect the complexity and dynamics of the bamboo gasification system or the uncertainty and variability of its environment. There may be some interactions or feedback that are not represented or accounted for in the conceptual model.
- It may not be applicable or generalizable to other regions or countries that have different conditions or contexts than Uganda. There may be some differences or variations that require adaptation or modification of the conceptual model.

In conclusion, this chapter has provided a conceptual and methodological framework for studying bamboo gasification as a social innovation for sustainable energy and rural development in Uganda. We have developed a conceptual model that captures the main elements and interactions of the bamboo gasification system and its environment. We have also proposed a 5-step approach that guides the decision makers to apply the conceptual model to any region or country that has bamboo resources. We have also discussed the research questions, hypotheses, data collection, and analysis methods that are derived from our conceptual model. Finally, we have acknowledged some limitations of our conceptual model and suggested some directions for future research.

**Acknowledgements** The author would like to thank Dr. Tabet, F. and his student, Nantongo, I., for their work on the earlier draft of this chapter, and the data collection was funded by PAUWES, University of Tlemcen, Algeria during master's thesis of Nantongo, I.

# References

African Development Bank Group (2016) SEFA Funds the preparation of the first-ever biomass gasification project in Uganda. https://www.afdb.org/en/news-and-events/sefa-funds-preparation-of-first-ever-biomass-gasification-project-in-uganda-16582

Africanews (2022) Ugandans produce alternative energy from bamboo. Retrieved from https://www.africanews.com/2022/03/28/ugandan-produces-alternative-energy-from-bamboo/

Antonini C, Treyer K, Moioli E, Bauer C, Schildhauer TJ, Mazzotti M (2016) Hydrogen from wood gasification with CCS—a techno-environmental analysis of production and use as a transport fuel Sustain Energy Fuels 4(10):5189–5208

Basu P, Borowski PF, Stępień P (2020a) Bamboo is an innovative biomass for the production of green energy by power plants Energies 15:1928

Basu P, Kaur J, Singh R, Kumar A (2020b) Bamboo biomass: A renewable source for bioenergy production Renew Energy Focus https://doi.org/10.1016/j.ref.2020.01.001

BMAU (2015) Biomass technology in Uganda: the unexploited energy potential. https://www.finance.go.ug/sites/default/files/Publications/BMAU%20Briefing%20Paper%205-15%20-%20Biomass%20Technology%20in%20Uganda.%20The%20Unexploited%20Energy%20Potential.pdf

Bridgwater AV (2012) Review of fast pyrolysis of biomass and product upgrading Biomass Bioenerg 38:68–94 https://doi.org/10.1016/j.biombioe.2011.01.048

Borowski G, Patuk I, Bandala ER (2021) Bamboo: Africa's untapped potential. Sustain 14(4):1955. https://www.iea.org/reports/energy-technology-perspectives-2020

Bucholz E, Da Silva I (2010) Energy access scenarios to 2030 for the power sector in sub-Saharan Africa. Utility Policy

CIF (2023) Uganda: forest investment program (FIP). https://www.climateinvestmentfunds.org/country/uganda/forest-investment-program-fip

Dincer I, Rosen MS (2012) Exergy: energy, environment and sustainable development Elsevier Science

Drift A, Boerrigter H, Coda B, Cieplik M, Hemmes K (2015) Challenges in biomass gasification. In: Proceedings of the International conference on polygeneration strategies (ICPS15). Vienna University of Technology

Eberhardt M, Foster V, Ouedraogo F, Santini M (2005) Access to energy services in Sub-Saharan Africa: priorities and challenges for the international community. https://www.researchgate.net/publication/23755413_Access_to_Energy_Services_in_Sub-Saharan_Africa_Priorities_and_Challenges_for_the_International_Community

Forest Machine Magazine (2020) Plant power: bamboo biomass in Uganda. Retrieved from https://forestmachinemagazine.com/plant-power-bamboo-biomass-in-uganda/

GlobeNewswire (2022) Biomass gasification global market report 2022. Retrieved from https://www.globenewswire.com/news-release/2022/09/29/2525567/0/en/Biomass-Gasification-Global-Market-Report-2022.html

Heritage Foundation (2023) Uganda's economy: population, GDP, inflation, business, trade, FDI, corruption. https://www.heritage.org/index/country/uganda

Hrbek J, Lundgren J, Waldheim L (2021) Emerging gasification technologies for waste & biomass. IEA Bioenergy: Task 33. Retrieved from https://www.ieabioenergy.com/wp-content/uploads/2021/02/Emerging-Gasification-Technologies_final.pdf

IEA (2020a) Global Energy Review 2020: The impacts of the Covid-19 crisis on global energy demand and $CO_2$ emissions. Paris: International Energy Agency. https://www.iea.org/reports/global-energy-review-2020

IEA (2020b) World Energy Outlook 2020. Paris: International Energy Agency. https://www.iea.org/reports/world-energy-outlook-2020

IEA (2022) Africa energy outlook 2022: key findings. https://www.iea.org/reports/africa-energy-outlook-2022/key-findings

INBAR (2018a) Bamboo for land restoration: policy synthesis report for Uganda. International Bamboo and Rattan Organisation

INBAR (2018b) National bamboo strategy and action plan for Uganda (2018–2022). https://www.inbar.int/wp-content/uploads/2020/05/1561628529.pdf

International Trade Administration (2022) Uganda—Energy. https://www.trade.gov/country-commercial-guides/uganda-energy

IRENA (2020) Uganda: energy country profile. https://www.irena.org/-/media/Files/IRENA/Agency/Statistics/Statistical_Profiles/Africa/Uganda_Africa_RE_SP.pdf

Jafri Y, Waldheim L, Lundgren J (2020) Status report on thermal gasification of biomass and waste 2019. IEA Bioenergy: Task 33. Retrieved from https://gasificationofbiomass.org/app/webroot/files/file/publications/special%20projects/2019/Status%20Report2019_final.pdf

Lobovikov M, Paudel S, Piazza M, Ren H, Wu J (2017a) Bamboo: the plant and its uses. Springer

Lobovikov M, Paudel S, Piazza M, Ren H, Wu J (2017b) Bamboo: an overlooked biomass resource? Biofuels Bioproducts & Biorefining-Biofpr 11(1):110–130

MacroTrends (2021) Uganda GDP per capita 1960–2023. https://www.macrotrends.net/countries/UGA/uganda/gdp-per-capita

Menin L, Vakalis S, Benedetti V, Patuzzi F, Baratieri M (2021a) Techno-economic assessment of integrated biomass gasification, electrolysis, and syngas methanation process Biomass Convers Biorefinery https://doi.org/10.1007/s13399-020-00654-9

Menin L, Sartori MMP, Souza AM, Souza OMMF, Rocha JD, Ronsani V (2021b) Biochar production from bamboo waste: A review on pyrolysis conditions optimization methods J Environ Manage https://doi.org/10.1016/j.jenvman.2020.111749

Monitor (2022) Bamboo growing: earn top dollar, save the environment. Retrieved from https://www.monitor.co.ug/uganda/magazines/farming/bamboo-growing-earn-top-dollar-save-the-environment-3721844

Mulgan G, Tucker S, Ali R, Sanders B (2007) Social innovation: what it is, why it matters, and how it can be accelerated. Skoll Centre for Social Entrepreneurship

Mwampamba TH, Kibwage JK, Mwakalukwa EE, Mwangi-Taylor E (2019) Bamboo bikes: a sustainable transport solution for Africa? The Guardian. Retrieved from https://www.theguardian.com/global-development/2019/dec/28/fighters-and-bamboo-bikes-the-african-innovators-driving-change

Mwampamba TH, Ghilardi A, Sander K, Chaix KJ (2021a) Bamboo biomass: a renewable energy alternative to charcoal production in sub-Saharan Africa? Renewable Energy 164:1219–1230

Mwampamba TH, Njenga-Bauerle M, Ochieng-Adimoa C, Njenga-Bauerle M (2021b) Bamboo for sustainable development in Africa: a review of the state of knowledge, challenges, and opportunities Sustainability 13(17):9911

Nantongo I (2017) Techno-economic feasibility of a gasification plant for rural electrification, under a bamboo-based sustainable economic model: case of Bududa district in Eastern Uganda. Accessed from http://repository.pauwes-cop.net/bitstream/handle/1/111/MT_Irene%20Nantongo.pdf?isAllowed=y&sequence=1

Off-Grid Solar Energy Market Assessment Brief—Uganda (2023) https://www.usaid.gov/sites/default/files/2022-12/Power-Africa-Market_Assessment-Brief-Uganda.pdf

Okello C, Pindozzi S, Faugno S, Boccia L (2013a) Development of bioenergy technologies in Uganda: a review of progress Renew Sustain Energy Rev https://doi.org/10.1016/j.rser.2012.11.021

Okello C, Pindozzi S, Faugno S, Boccia L (2013b) Development of bioenergy technologies in Uganda: A review of progress Renew Sustain Energy Rev 18:55–63

Partey ST, Kwaku M, Buerkert A, Ofori DA (2017a) Potential role of bamboo in alleviating poverty: insights from Ghanaian smallholder farmers' preference for land use options incorporating bamboo production systems in southern Ghana Land Use Policy https://doi.org/10.1016/j.landusepol.2017.08.027

Partey ST, Sarfo DA, Frith O, Kwaku M, Thevathasan NV (2017b) Potentials of bamboo-based agroforestry for sustainable development in Sub-Saharan Africa: a review Agric Res 6:22–32

Phills JA, Deiglmeier K, Miller DT (2008) Rediscovering social innovation Stanf Soc Innov Rev 6(4):34–43

Powermax Gasifiers (2023) Bamboo waste, bamboo chips gasification power plant-biomass gasification power generation system-Biomass Gasifier|Biomass Gasification Power Plant|Wuxi Teneng Power Machinery Co., Ltd. Retrieved from https://www.powermaxgasifiers.com/index.php?ac=article&at=read&did=260

SEforALL Africa Hub (2023) Uganda: at a glance. https://www.se4all-africa.org/seforall-in-africa/country-data/uganda/

Sikarwar VS, Zhao M, Clough P, Yao J, Zhong X, Memon MZ, Shah N, Anthony EJ, Fennell PS (2016) An overview of advances in biomass gasification Energy Environ Sci 9(10):2939–2977

The Independent (2023) Ugandan bamboo entrepreneur paves the way for sustainable energy solutions. https://www.independent.co.ug/ugandan-bamboo-entrepreneur-paves-way-for-sustainable-energy-solutions/

Truong AH, Le TMA (2014) Overview of bamboo biomass for energy production. Retrieved from https://shs.hal.science/halshs-01100209/document

United Nations Africa Renewal Magazine (2016) African democracy coming of age. Africa Renewal, August–November 2016. https://prod.iea.org/reports/world-energy-investment-2020?mode=overview

U.S. Department of Energy (2023) Hydrogen production: biomass gasification. Retrieved from https://www.energy.gov/eere/fuelcells/hydrogen-production-biomass-gasification

Waldheim L, Hrbek J, Lundgren J (2022) Status report on thermal gasification of biomass and waste 2021. IEA Bioenergy: Task 33

Wikipedia (2021) Geography of Uganda. https://en.wikipedia.org/wiki/Geography_of_Uganda

WorldAtlas (2020) What type of climate does Uganda have? https://www.worldatlas.com/articles/what-type-of-climate-does-uganda-have.html

World Bank (2020a) World development indicators: Uganda. https://databank.worldbank.org/reports.aspx?source=2&country=UGA

World Bank (2020b) Surface area (sq. km)—Uganda. https://data.worldbank.org/indicator/AG.SRF.TOTL.K2?locations=UG

World Bank (2020c) Uganda overview: development news, research, and data. https://www.worldbank.org/en/country/uganda/overview.

World Bank (2021) World Development Report 2021: Data for Better Lives. Washington, DC: World Bank. https://www.iea.org/reports/world-energy-investment-2020

World Bank (2023) Access to electricity, urban (% of urban population)—Uganda. https://data.worldbank.org/indicator/EG.ELC.ACCS.UR.ZS?locations=UG

Zafar S, Kumar A, Kumar S (2017) Global challenges in the sustainable development of biomass gasification: an overview Renew Sustain Energy Rev 80 23 43

# Chapter 5
# Renewable Energy and Sustainable Development in South Africa: Challenges, Barriers and Solutions

**Abstract** This chapter explores how renewable energy can support sustainable development in South Africa. It reviews the literature on four topics: the current and future trends of renewable energy use and production; the factors that influence renewable energy adoption and diffusion; the effects of renewable energy on different aspects of sustainability; and the policy and technical measures to enhance the role of renewable energy in achieving sustainability goals. The chapter shows that South Africa has made considerable progress in using and producing renewable energy, but still faces some obstacles and challenges that need to be addressed. The chapter proposes some potential policy and technical solutions to create a favorable environment for renewable energy transition in South Africa. The chapter also acknowledges some research gaps and limitations and suggests some areas for further research. The chapter aims to enrich the knowledge of renewable energy and sustainability and provide useful insights and lessons for policymakers, investors, developers, and users of renewable energy technologies in South Africa and other countries.

*What is in for the readers of this chapter*? In this chapter, researchers can avail a comprehensive and updated review of the literature on renewable energy and sustainable development in South Africa, as well as identify some research gaps and limitations, and directions for further research. Policymakers can gain valuable insights and lessons on the policy and technical solutions for enhancing the role of renewable energy in advancing sustainable development in South Africa and beyond. Practitioners can learn about the current status and trends of renewable energy development and consumption in South Africa, as well as the drivers and barriers to renewable energy adoption and diffusion. University's research libraries and research centers can access useful references or resources on renewable energy and sustainable development issues. Students both undergraduate and graduate can enhance their knowledge and understanding of the impacts of renewable energy on various dimensions of sustainable development in South Africa, as well as the trade-offs and synergies between renewable energy and other development goals and priorities.

© The Author(s), under exclusive license to Springer Nature Switzerland AG 2024     89
H. Qudrat-Ullah, *Exploring the Dynamics of Renewable Energy and Sustainable Development in Africa*, Advances in African Economic, Social and Political Development, https://doi.org/10.1007/978-3-031-48528-2_5

## 5.1  Introduction

Renewable energy (RE) and sustainable development are two concepts that are closely linked and have become globally important in recent years. RE refers to the energy sources that can be naturally replenished and have little or no carbon emissions, such as solar, wind, hydroelectric, and biomass. Sustainable development refers to the development that satisfies the present needs without jeopardizing future needs, considering the environmental, social, and economic aspects. The use of RE can support sustainable development by lowering greenhouse gas emissions, enhancing energy security, creating jobs, and improving human well-being.

Renewable energy (RE) is energy that comes from natural resources that can be naturally restored, such as solar, wind, hydro, biomass, and geothermal. RE can help achieve sustainable development, which is the development that balances the needs of the present and the future in terms of environmental, social, and economic aspects. Some of the benefits of RE for sustainable development are: reducing greenhouse gas emissions that cause climate change, enhancing energy security by diversifying energy sources, creating jobs and income opportunities, improving health by reducing air pollution, and alleviating poverty by providing affordable and clean energy (IRENA 2020a, b, c). However, the adoption and diffusion of RE technologies (RETs) face various drivers and barriers that affect their viability and impact in different contexts. This literature review aims to identify the drivers and barriers to RE adoption and diffusion in South Africa and analyze how they affect sustainability outcomes.

South Africa is the third-largest economy and the highest primary energy consumer in Africa. The country's energy sector is dominated by coal, which accounts for about 90% of electricity generation and 70% of primary energy supply. Coal is also the main source of greenhouse gas emissions and air pollution in the country, contributing to climate change and health problems.

According to the literature, the current energy supply mix of South Africa is dominated by coal, which accounts for about 90% of electricity generation and 70% of the primary energy supply (Li et al. 2020a, b, c, d; Nhamo and Nhamo 2019b). Coal is also the main source of greenhouse gas emissions and air pollution in the country, contributing to climate change and health problems (Li et al. 2020a, b, c, d). The other sources of energy supply in South Africa are crude oil (21.6%), renewable and wastes (7.6%), gas (2.8%), nuclear (0.4%), and hydro (0.1%) (Energypedia 2023). RE sources include solar, wind, hydro, biomass, and geothermal (Our World in Data 2023). RE accounted for 11.6% of the total installed capacity and 6.25% of the total electricity generation in 2019 (DMRE 2021).

South Africa is a country that has great potential for RE development and a strong commitment to sustainable development. According to the International RE Agency (IRENA), South Africa is the second-largest RE market in Africa, after Egypt, and has successfully scaled up national renewables-based generation at a competitive cost (IRENA 2020a, b, c). South Africa has committed to reducing its emissions by 28% by 2030 under the Paris Agreement and has developed ambitious plans to gradually phase out coal and increase the share of RE in its electricity mix (Li et al. 2020a, b, c, d). However, South Africa still faces many challenges and barriers in achieving its RE and sustainable development goals, such as the high cost of RE technologies, the regulatory environment, the integration of embedded generation, and the socio-economic impacts of the energy transformation (IRENA 2020a, b, c).

The purpose of this study is to explore the relationship between RE and sustainable development in South Africa and to provide recommendations for improving the potential of RE for achieving sustainable development goals. The specific objectives of this study are:

- To assess the current status and trends of RE development and consumption in South Africa and compare it with other countries in the region and globally.
- To identify the drivers and barriers to RE adoption and diffusion in South Africa and analyze how they affect the sustainability outcomes.
- To evaluate the impacts of RE on various dimensions of sustainable development, such as greenhouse gas emissions, energy security, economic growth, job creation, poverty reduction, and human well-being.
- To propose policy and technical solutions for enhancing the role of RE in advancing sustainable development in South Africa and overcoming the existing challenges and barriers.

The methodology of this study will consist of a literature review. This study will contribute to the existing knowledge on RE and sustainable development by providing empirical evidence and a comprehensive analysis of the South African context. It will also provide practical implications and recommendations for policy-makers, practitioners, researchers, and stakeholders who are interested in promoting RE and sustainable development in South Africa and beyond.

The rest of this chapter is organized as follows: Sect. 5.2 describes the methodology of this study. Section 5.3 reviews the current status and trends of RE development and consumption in South Africa and compares it with other countries in the region and globally; identifies the drivers and barriers for RE adoption and diffusion in South Africa and analyzes how they affect the sustainability outcomes; evaluates the impacts of RE on various dimensions of sustainable development, such as greenhouse gas emissions, energy security, economic growth, job creation, poverty reduction, and human well-being; and proposes policy and technical solutions for enhancing the role of RE in advancing sustainable development in South Africa and overcoming the existing challenges and barriers. Section 5.4 concludes the chapter with a discussion and summary of the main findings and recommendations and some directions for further research.

## 5.2   Methodology

This study aims to analyze the policy and technical solutions for enhancing the role of RE in advancing sustainable development in South Africa. The study adopts a literature review approach that synthesizes and evaluates the existing literature on the topic. The study searches for relevant literature from various sources, such as academic journals, books, reports, websites, etc., using keywords and criteria related to the topic. The study selects the literature based on the quality, relevance, and currency of the sources. The study analyzes the selected literature using thematic analysis, which is a method of identifying and organizing the main themes and issues emerging from the literature. The study uses a deductive approach to derive the themes from the research questions and objectives of the study.

## 5.3   Literature Review and Conceptual Model Building

South Africa has abundant RE resources, especially solar and wind, which are among the most competitive in the world. The country has successfully scaled up RE generation through the RE Independent Power Producer Procurement Programme (REIPPPP), which has attracted significant private-sector investment and created thousands of jobs. The REIPPPP has also contributed to reducing electricity costs, improving grid stability, and increasing access to electricity for rural communities (IRENA 2020a, b, c).

This section reviews the existing literature on RE development and consumption in South Africa and other countries in the region and globally. Consistent with the focus of this study, the literature review covers four main aspects: (1) *the current status and trends of RE development and consumption*; (2) *the drivers and barriers for RE adoption and diffusion*; (3) *the impacts of RE on various dimensions of sustainable development*; and (4) *the policy and technical solutions for enhancing the role of RE in advancing sustainable development*.

### 5.3.1   Current Status and Trends of RE Development and Consumption

Renewable energy (RE) is energy that comes from natural resources that can be naturally restored, such as solar, wind, hydroelectric, and biomass. RE produces little or no carbon emissions that cause climate change. RE made up 26.4% of the world's electricity production in 2018, which was an increase from 23.3% in 2015, and is expected to reach 30% by 2024 (IEA 2019). The main reasons for the growth of RE are the lower costs of RE technologies, the higher concerns about climate change and air pollution, the higher demand for electricity access and security, and the supportive policies and regulations in many countries (IRENA 2020a, b, c).

South Africa is a leading country in Africa in using and producing RE. RE is energy that comes from natural resources that can be naturally restored, such as solar, wind, hydroelectric, and biomass. RE produces little or no carbon emissions that cause climate change. By the end of 2019, South Africa had 6,422 MW of RE capacity, which was 11.6% of its total capacity of 55,247 MW (DMRE 2021). The main types of RE in South Africa were solar PV, wind, hydroelectric, and biomass. Solar PV made up 42.8% of the RE capacity, followed by wind (41.7%), hydroelectric (12.5%), and biomass (3%) (DMRE 2021). In 2019, South Africa produced 15,316 GWh of electricity from RE, which was 6.25% of its total electricity production of 244,843 GWh (DMRE 2021). The share of RE in electricity production increased from 2.5% in 2015 to 6.25% in 2019, showing a fast growth of RE in South Africa (DMRE 2021).

South Africa's RE development and consumption can be compared with other countries in the region and globally. According to IRENA (2020a, b, c), South Africa was the second-largest RE market in Africa, after Egypt, and ranked 22nd in the world in terms of installed RE capacity in 2019. South Africa also had the highest share of variable RE (VRE), such as solar PV and wind, in its electricity mix among African countries, reaching 8.7% in 2019 (IRENA 2020a). However, South Africa still lagged behind some of the global leaders in RE development and consumption, such as China, Germany, Brazil, India, etc., which had higher installed capacities, generation shares, and VRE penetration levels of RE than South Africa (IRENA 2020a).

Figure 5.1 presents our conceptual model for RE development and consumption in South Africa. The dependent variable in this model is the RE development and consumption in South Africa, which is measured by the installed capacity, generation share, and VRE penetration level of RE sources. The independent variables are the factors that influence RE development and consumption in South Africa, such as the costs of RE technologies, the concerns about climate change and air pollution, the demand for electricity access and security, and the supportive policies and regulations in the country. If we want to compare South Africa with other countries in terms of RE development and consumption, then it can be considered as a control variable.

In conclusion, this section has reviewed the current status and trends of RE development and consumption in South Africa and compared them with other countries in the region and globally. The review has shown that South Africa has made significant progress in RE development and consumption in recent years, driven by various factors such as cost reduction, policy support, climate action, and energy security. However, the review has also revealed that South Africa still faces some challenges and barriers that limit the potential and impact of RE in advancing sustainable development, such as policy uncertainty, grid constraints, financing gaps, and social resistance. Therefore, using our conceptual model further empirical research is needed to identify and analyze the drivers and barriers for RE adoption and diffusion in South Africa and to propose policy and technical solutions for enhancing the role of RE in advancing sustainable development.

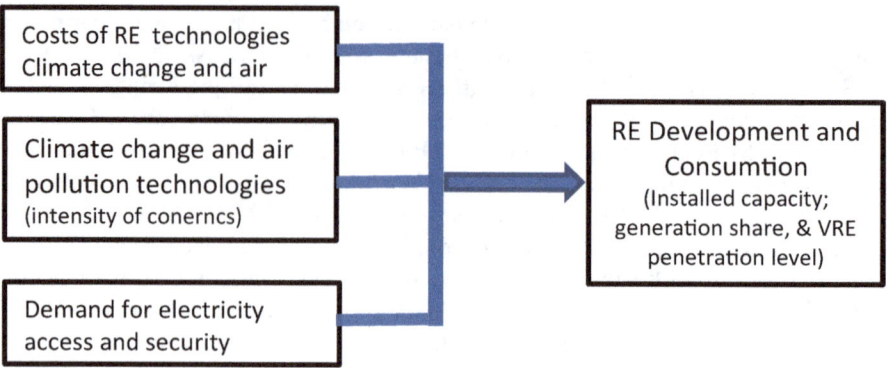

**Fig. 5.1**  The conceptual model of RE development and consumption in South Africa

## 5.3.2   Drivers and Barriers to RE Adoption and Diffusion

RE adoption and diffusion are influenced by various factors that can be classified into four categories: technical, economic, social, and institutional (Painuly et al. 2018). Technical factors refer to the availability and quality of RE resources, the performance, and reliability of RE technologies, the integration and compatibility of RE with existing infrastructure and systems, etc. Economic factors refer to the costs and benefits of RE production and consumption, the availability and accessibility of financing options for RE projects, the competitiveness and attractiveness of RE compared to conventional energy sources, etc. Social factors refer to the awareness and acceptance of RE among different stakeholders, such as consumers, and producers. Some of these barriers include:

- Policy and legal barriers: The lack of clear and consistent policy frameworks and regulations for RE development, integration, and procurement; the uncertainty and delays in policy implementation and decision making; the vested interests and resistance from the coal industry and labor unions; the inadequate alignment of RE policies with other national policies and strategies for economic development, industrialization, job creation, and social justice (Murombo 2022a; IRENA 2020a, b, c).
- Technical barriers: The limited capacity and reliability of the grid infrastructure to accommodate variable RE sources; the lack of adequate transmission and distribution networks to connect remote RE projects to the grid or off-grid customers; the insufficient technical skills and expertise for RE planning, installation, operation, and maintenance; the lack of standardized quality control and certification systems for RE equipment and services (IRENA 2020a, b, c; Nhamo and Nhamo 2019b).
- Financial barriers: The high upfront capital costs and risks associated with RE projects; the difficulty in accessing affordable and long-term financing from local banks and financial institutions; the competition from subsidized fossil fuels that

distort the market prices and incentives for RE; the lack of adequate fiscal and financial incentives and mechanisms to support RE development and deployment (IRENA 2020a, b, c; Nhamo and Nhamo 2019b).

These barriers need to be addressed through a combination of policy, regulatory, institutional, technical, financial, and social interventions that can create an enabling environment for RE adoption and diffusion in South Africa. When designing interventions to promote RE, it is also important to consider how RE affects other aspects of sustainability, such as economic growth, social equity, environmental protection, and climate resilience, and how these aspects affect RE in return. A holistic and systemic approach is needed to ensure that RE contributes to a just and inclusive transition to a low-carbon economy in South Africa.

This section has reviewed the existing literature on the drivers and barriers to RE adoption and diffusion in South Africa. The literature has identified various technical, economic, social, and institutional factors that influence the development and consumption of RE in the country. The literature has also suggested some possible interventions and solutions to overcome these factors and create a conducive environment for RE transition. However, the literature also reveals some research gaps and limitations that need to be addressed in future studies. Some of these gaps include:

- The lack of empirical evidence and data on the actual impacts and outcomes of RE policies and programs in South Africa, especially on the social and environmental dimensions of sustainable development.
- The lack of comprehensive and comparative analysis of the drivers and barriers for different types of RE technologies, such as solar, wind, hydro, biomass, etc., and their suitability and applicability for different contexts and locations in South Africa.
- The lack of participatory and inclusive research methods that involve the perspectives and experiences of various stakeholders, especially the local communities and groups that are affected by or benefit from RE projects.
- The lack of interdisciplinary and transdisciplinary research approaches that integrate the natural, social, and engineering sciences to address the complex and interrelated challenges and opportunities of RE transition in South Africa.

In our model, as shown in Fig. 5.2, the dependent variable is the adoption and diffusion of RE technologies, which is the outcome that the project aims to achieve. The independent variables are the drivers and barriers for RE adoption and diffusion, which are classified into four categories: technical, economic, social, and institutional. These factors can influence the development and consumption of RE in South Africa.

These research gaps indicate the need for more rigorous and comprehensive studies on RE adoption and diffusion in South Africa that can inform and support policy-making and practice in this field. Such studies can also contribute to the global knowledge base on RE and sustainable development and provide valuable insights and lessons for other countries and regions that are undergoing or planning to undergo similar transitions.

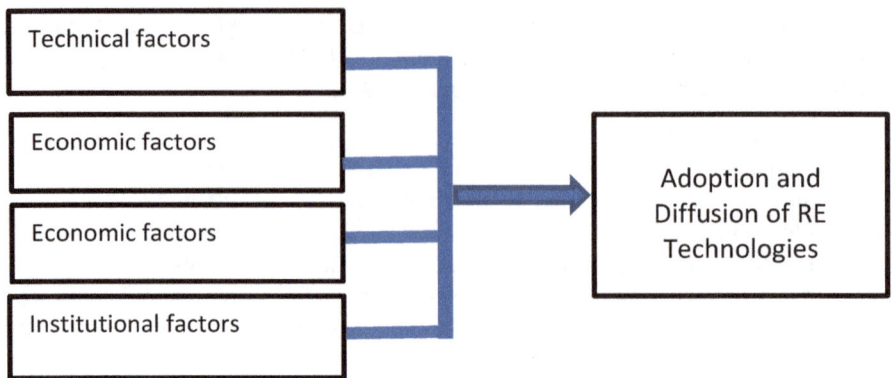

**Fig. 5.2**  The conceptual model of adoption and diffusion of RE technologies in South Africa

### 5.3.3  The Impacts of RE on Various Dimensions of Sustainable Development

This section explores how RE affects different aspects of sustainability. Sustainability is the development that balances the needs of the present and the future in terms of environmental, social, and economic aspects (United Nations, 1987). RE can support sustainability by lowering greenhouse gas emissions that cause climate change, enhancing energy security by diversifying energy sources, creating jobs and income opportunities, and improving human well-being by reducing air pollution and providing affordable and clean energy. However, RE also involves some trade-offs and challenges that need to be considered and managed. The section reviews the literature on four main aspects of sustainability: (1) environmental; (2) economic; (3) social; and (4) institutional.

*Environmental Dimension.* The environmental dimension of sustainable development refers to the protection and conservation of natural resources and ecosystems, as well as the mitigation and adaptation to climate change and other environmental risks. RE can have positive impacts on the environmental dimension by reducing greenhouse gas emissions and air pollution, which are the main drivers of climate change and its adverse effects on human health and biodiversity. According to IRENA (2019), RE sources avoided 2 gigatons of carbon dioxide equivalent (GtCO2eq) emissions in 2018, equivalent to 7% of total energy-related CO2 emissions. RE can also reduce water consumption and land use compared to conventional energy sources, which can have significant implications for water scarcity and food security (IRENA 2019).

However, RE also has some negative impacts on the environmental dimension that need to be considered and minimized. For instance, RE technologies require materials and resources for their production, installation, operation, and disposal, which can generate waste and pollution along their life cycle. RE projects can also affect local ecosystems and biodiversity through land use change, habitat loss or fragmentation, noise or visual disturbance, collision or displacement of wildlife, etc. (Hernandez et al. 2020). Therefore, RE development needs to be planned and implemented in a way that avoids or mitigates these impacts through appropriate site selection, design, monitoring, and compensation measures.

*Economic Dimension.* The economic dimension of sustainable development refers to the promotion of economic growth, productivity, competitiveness, innovation, and diversification, as well as the reduction of poverty, inequality, and vulnerability. RE can have positive impacts on the economic dimension by creating new markets, industries, jobs, and income opportunities along the RE value chain. According to IRENA (2020a, b, c), RE employed 11.5 million people worldwide in 2019, with solar photovoltaic being the largest employer with 3.8 million jobs. RE can also reduce the dependence on fossil fuel imports and enhance the energy security and resilience of countries and regions. RE can also lower the costs of electricity generation and consumption by reducing fuel costs, operation, and maintenance costs, transmission and distribution losses, etc. (IRENA 2019).

However, RE also faces some challenges and barriers in the economic dimension that need to be overcome and removed. For instance, RE technologies often have high upfront capital costs that require adequate financing mechanisms and incentives to attract investors and consumers. RE projects can also face market distortions and competition from subsidized fossil fuels that undermine their profitability and viability. RE can also have distributional impacts on different sectors and groups that need to be assessed and addressed. For example, RE can affect the revenues and employment of the fossil fuel industry and its related sectors, which can create social and political resistance and opposition. RE can also create winners and losers among different regions, communities, and households, depending on their access to and affordability of RE services. Therefore, RE development needs to be accompanied by supportive policies and measures that address these challenges and barriers and ensure a fair and inclusive transition to a low-carbon economy.

*Social Dimension.* The social dimension of sustainable development refers to the improvement of human health, education, culture, gender equality, human rights, and social cohesion and participation. RE can have positive impacts on the social dimension by providing access to modern energy services for the poorest and most marginalized segments of society, especially in rural and remote areas. According to IRENA (2019), RE solutions, such as off-grid solar systems, mini-grids, and clean cooking devices, can provide affordable, reliable, and sustainable energy for lighting, communication, education, health care, income generation, etc. RE can also improve human health and well-being by reducing air pollution and its associated diseases and deaths. According to WHO (2018), household air pollution from burning solid fuels for cooking and heating caused 3.8 million premature deaths in 2016, mostly in low- and middle-income countries. RE can also empower women and girls by reducing their drudgery and time spent on collecting fuelwood and water, improving their safety and security, enhancing their education and employment opportunities, etc. (IRENA 2019).

However, RE also entails some social challenges and risks that need to be considered and mitigated. For instance, RE projects can affect the livelihoods, rights, and interests of local communities and indigenous peoples who depend on natural resources for their survival and identity. RE projects can also generate conflicts and disputes over land ownership, resource access, benefit sharing, etc. among different stakeholders. RE projects can also face social acceptance and resistance issues due to cultural, religious, or aesthetic reasons. Therefore, RE development needs to involve meaningful consultation and participation of all relevant stakeholders, especially the affected communities and groups, and ensure that they are informed, consulted, compensated, and benefited from the RE projects.

*Institutional Dimension.* The institutional dimension of sustainable development refers to the establishment and improvement of governance structures, policies, regulations, standards, norms, and practices that enable and facilitate the transition to RE. RE can have positive impacts on the institutional dimension by fostering cooperation and coordination among different actors and sectors at various levels (local, national, regional, and global) involved in RE development and consumption. RE can also stimulate innovation and learning by creating new knowledge, skills, capacities, networks, etc. that can enhance the performance and efficiency of RE systems.

This section has reviewed the existing literature on the impacts of RE on various dimensions of sustainable development. The literature has shown that RE can have positive impacts on the environmental, economic, social, and institutional dimensions by reducing greenhouse gas emissions and air pollution, enhancing energy security and access, creating jobs and income opportunities, improving human health and well-being, fostering cooperation and innovation, and promoting transparency and accountability. However, the literature has also highlighted some trade-offs and challenges that RE entails for sustainable development, such as waste and pollution generation, ecosystem and biodiversity impacts, financing and market barriers, distributional and social justice issues, livelihoods and rights conflicts, social acceptance, and resistance issues, etc. The literature has also suggested some possible interventions and solutions to address these trade-offs and challenges and ensure a balanced and holistic approach to RE transition.

However, the literature also reveals some research gaps and limitations that need to be addressed in future studies. Some of these gaps include:

- The lack of comprehensive and integrated assessment frameworks and indicators that can capture the multiple and interrelated impacts of RE on sustainable development across different scales, sectors, and contexts.
- The lack of empirical evidence and data on the long-term and cumulative impacts of RE on sustainable development, especially on the environmental and social dimensions, as well as the feedback loops and interactions between different dimensions.
- The lack of comparative analysis of the impacts of different types of RE technologies, such as solar, wind, hydro, biomass, etc., on sustainable development, taking into account their technical, economic, social, and institutional characteristics and implications.
- The lack of participatory and inclusive research methods that involve the perspectives and experiences of various stakeholders, especially the local communities

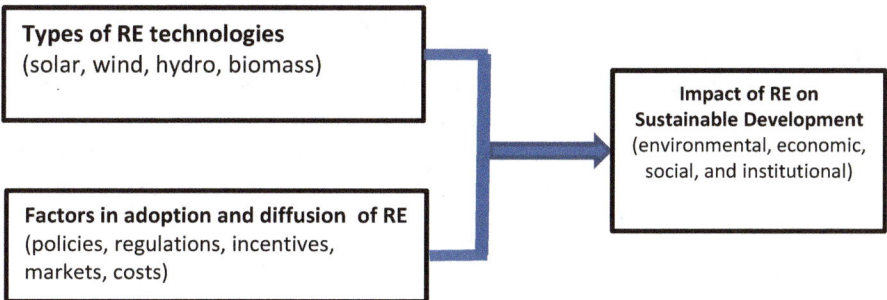

**Fig. 5.3** The conceptual model of the impact of RE on sustainable development in South Africa

and groups that are affected by or benefit from RE projects, in assessing the impacts of RE on sustainable development.

- The lack of policy analysis and evaluation studies that examine the effectiveness and efficiency of existing policies and programs that support or hinder RE transition concerning sustainable development goals and targets.

Figure 5.3 displays our conceptual model where the dependent variable is the impact of RE on sustainable development, which is measured by various indicators across four dimensions: environmental, economic, social, and institutional. The independent variables are the different types of RE technologies, such as solar, wind, hydro, biomass, etc., and the various factors that influence their adoption and diffusion, such as policies, regulations, incentives, markets, costs, etc.

These research gaps indicate the need for more rigorous theoretical and empirical studies on the impacts of RE on sustainable development that can inform and support policy-making and practice in this field. Such studies, using our conceptual model, as is shown in Fig. 5.3, while addressing the RE issues of South Africa, can also contribute to the global knowledge base on RE and sustainable development and provide valuable insights and lessons for other countries and regions that are undergoing or planning to undergo similar transitions.

### 5.3.4 The Policy and Technical Solutions

RE is energy that comes from natural resources that can be naturally restored, such as solar, wind, hydro, biomass, and geothermal. RE can help achieve sustainable development, which is the development that balances the needs of the present and the future in terms of environmental, social, and economic aspects (WCED 1987). These three aspects of sustainability are interrelated and affect each other. RE can contribute to each of these dimensions by providing multiple benefits, such as:

- Improving energy access: RE technologies can provide affordable, reliable, and clean energy services to people who lack access to modern energy sources, especially in rural and remote areas. By enhancing energy access, RE can improve living standards, health, education, and gender equality (IRENA 2020a, b, c).
- RE technologies can help countries become more energy secure by using their resources instead of importing fuels from other countries and by making their energy supply more reliable. For example, countries can use solar and wind power, which are resources that are available locally and do not run out, to reduce their reliance on foreign energy sources and to have more control over their energy supply (Karekezi et al. 2003).
- Reducing greenhouse gas emissions: RE technologies can reduce greenhouse gas emissions by displacing fossil fuels in power generation, heating, cooling, and transport. By mitigating climate change, RE can protect the environment and human health from the adverse impacts of global warming (Li et al. 2020a, b, c, d).
- Creating jobs and income: RE technologies can create jobs and income opportunities along the value chain, from manufacturing and installation to operation and maintenance. By stimulating economic growth and social inclusion, RE can reduce poverty and inequality (IRENA 2020a, b, c).

However, the adoption and diffusion of RE technologies also face various challenges and barriers that limit their potential and impact in different contexts. These challenges and barriers can be classified into two broad categories: policy and technical. Policy challenges and barriers refer to the lack of clear and consistent policy frameworks and regulations that support RE development and deployment. Policy challenges and barriers can include:

- Inadequate or conflicting policies and targets for RE at different levels of governance (national, regional, local) (IRENA 2020a, b, c).
- Lack of coordination and integration of RE policies with other relevant policies and strategies, such as economic development, industrialization, job creation, social justice, climate change mitigation, and adaptation (Murombo 2022b).
- Vested interests and resistance from incumbent fossil fuel industries and actors that influence policy-making and implementation (Handler et al. 2021).
- Insufficient or distorted incentives and mechanisms to promote RE investment and innovation, such as subsidies, tariffs, taxes, feed-in tariffs, auctions, and quotas (IRENA 2020a, b, c).

Technical challenges and barriers refer to the lack of adequate infrastructure and capacity to accommodate RE technologies in the energy system. Technical challenges and barriers can include:

- Limited capacity and reliability of the grid infrastructure to integrate variable RE sources without compromising stability and quality of supply (IRENA 2020a, b, c).
- Lack of adequate transmission and distribution (T&D) networks to connect RE projects to the grid or off-grid customers (Karekezi et al. 2003).

- Insufficient technical skills and expertise for RE planning, installation, operation, and maintenance (Nhamo and Nhamo 2019a).
- Lack of standardized quality control and certification systems for RE equipment and services (Karekezi et al. 2003).

To overcome these challenges and barriers, various policy and technical solutions have been proposed and implemented in different contexts. Policy solutions aim to create an enabling environment for RE adoption and diffusion by providing clear and consistent policy frameworks and regulations that support RE development and deployment. Policy solutions can include:

- Developing or revising national or regional RE policies or strategies that set ambitious but realistic targets for RE share in the energy mix (IRENA 2020a, b, c).
- Aligning or integrating RE policies with other relevant policies or strategies that address economic development, industrialization, job creation, social justice, climate change mitigation, and adaptation (Murombo 2022a, b).
- Engaging or involving various stakeholders in policy making and implementation process, such as government agencies, the private sector, civil society, academia, and local communities (Handler et al. 2021).
- Providing or enhancing fiscal and financial incentives and mechanisms to promote RE investment and innovation, such as subsidies, tariffs, taxes, feed-in tariffs, auctions, and quotas (IRENA 2020a, b, c).

Technical solutions aim to improve the infrastructure and capacity to accommodate RE technologies in the energy system. Technical solutions can include:

- Expanding or upgrading the grid infrastructure to integrate variable RE sources without compromising stability and quality of supply (IRENA 2020a, b, c).
- Developing or extending transmission and distribution networks to connect RE projects to the grid or off-grid customers (Karekezi et al. 2003).
- Building or enhancing technical skills and expertise for RE planning, installation, operation, and maintenance (Nhamo and Nhamo 2019a).
- Establishing or improving standardized quality control and certification systems for RE equipment and services (Karekezi et al. 2003).

In the conceptual model, as shown in Fig. 5.4, the dependent variable is the adoption and diffusion of RE technologies, which is the outcome that the project aims to achieve. The independent variables are the policy and technical solutions that are proposed and implemented to overcome the challenges and barriers to RE development and deployment. These solutions can influence the adoption and diffusion of RE technologies in different contexts.

These policy and technical solutions are not mutually exclusive, but rather complementary and interdependent. A holistic and systemic approach is needed to ensure that RE technologies can play an effective and efficient role in advancing sustainable development in different contexts.

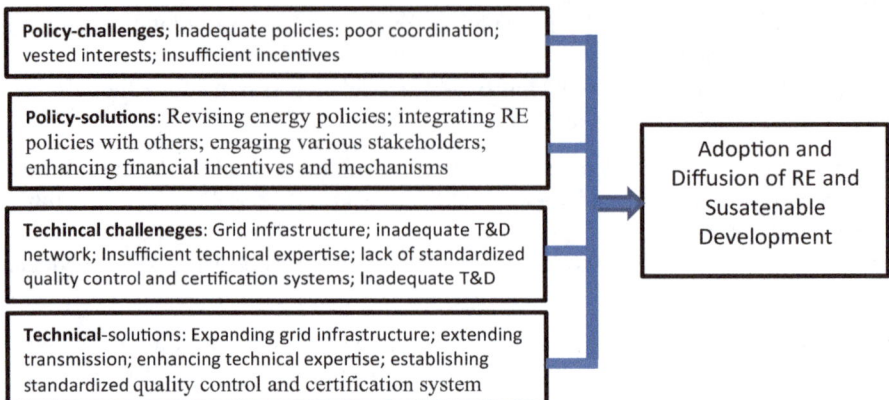

**Fig. 5.4** The conceptual model of adoption and diffusion RE and sustainable development

## 5.3.5  *Towards a Low-Carbon and Sustainable Future in South Africa*

South Africa is facing a critical challenge of transforming its energy sector to achieve a low-carbon and sustainable future. What could be a way forward a low-carbon and sustainable development in South Africa? Our literature analysis provides some actional steps that policymakers can embrace:

- South Africa should gradually phase out coal and increase the share of RE in its electricity mix, as coal is the main source of greenhouse gas emissions and air pollution in the country, contributing to climate change and health problems (Li et al. 2020a, b, c, d; Nhamo and Nhamo 2019a, b, c). Coal is also becoming less competitive and more risky as a source of energy, due to the declining costs of RE technologies, the increasing carbon taxes and regulations, and the growing social and environmental awareness and activism (IRENA 2020a, b, c; Murombo 2022b).
- South Africa should aim to achieve around 35% to 40% of renewable electricity in its supply mix by 2030, corresponding to 30 gigawatts (GW) of RE capacity, up from only 9.6 GW in 2020 (IRENA Coalition for Action 2021). This target is consistent with South Africa's Low Emission Development Strategy of 2020 and its commitment to reduce its emissions by 28% by 2030 under the Paris Agreement (Li et al. 2020a, b, c, d; IRENA Coalition for Action 2021).
- South Africa should focus on developing wind and solar power as the main sources of RE, as they have the highest potential and availability in the country (Our World in Data 2023; Murombo 2022a). Wind and solar power can also provide multiple benefits for sustainable development, such as reducing greenhouse gas emissions, enhancing energy security, creating jobs, and improving human well-being (IRENA 2020a, b, c; Our World in Data 2023).

• South Africa should implement policy and technical solutions to overcome the existing challenges and barriers to RE adoption and diffusion, such as policy uncertainty, grid constraints, financing gaps, and social resistance (IRENA 2020a, b, c; Nhamo and Nhamo 2019a, b, c; Murombo, 2022b). Some of these solutions include developing or revising national or regional RE policies or strategies that set ambitious but realistic targets for RE share in the energy mix; aligning or integrating RE policies with other relevant policies or strategies that address economic development, industrialization, job creation, social justice, climate change mitigation and adaptation; engaging or involving various stakeholders in policy making and implementation process, such as government agencies, private sector, civil society, academia, local communities; providing or enhancing fiscal and financial incentives and mechanisms to promote RE investment and innovation, such as subsidies, tariffs, taxes, feed-in tariffs, auctions, quotas; expanding or upgrading the grid infrastructure to integrate variable RE sources without compromising stability and quality of supply; developing or extending transmission and distribution networks to connect RE projects to the grid or to off-grid customers; building or enhancing technical skills and expertise for RE planning, installation, operation and maintenance; establishing or improving standardized quality control and certification systems for RE equipment and services (IRENA 2020a, b, c).

In conclusion, this section has discussed the policy and technical solutions for enhancing the role of RE in advancing sustainable development. The section has shown that various policy and technical solutions have been proposed and implemented in different contexts to overcome the challenges and barriers that hinder RE adoption and diffusion. However, the section has also revealed that there is still a lack of empirical evidence and analysis on the effectiveness and impact of these solutions in different contexts and scenarios. Therefore, further research is needed to evaluate and compare the policy and technical solutions for RE in terms of their costs, benefits, risks, trade-offs, and synergies for sustainable development on a regional and global basis. Such research can provide valuable insights and recommendations for policymakers, investors, developers, and users of RE technologies.

## 5.4  Discussion and Conclusion

This study has reviewed the existing literature on RE development and consumption in South Africa and other countries in the region and globally. The review has shown that South Africa has made significant progress in RE development and consumption in recent years, driven by various factors such as cost reduction, policy support, climate action, and energy security. However, the review has also revealed that South Africa still faces some challenges and barriers that limit the potential and impact of RE in advancing sustainable development, such as policy uncertainty, grid constraints, financing gaps, and social resistance. Therefore, further research is needed to identify and analyze the drivers and barriers to RE adoption and diffusion in South Africa and to propose policy and technical solutions for enhancing the role of RE in advancing sustainable development.

The review has also highlighted some of the impacts of RE on various dimensions of sustainable development, such as environmental, economic, social, and institutional. The review has shown that RE can have positive impacts on these dimensions by reducing greenhouse gas emissions and air pollution, enhancing energy security and access, creating jobs and income opportunities, improving human health and well-being, fostering cooperation and innovation, and promoting transparency and accountability. However, the review has also indicated some trade-offs and challenges that RE entails for sustainable development, such as waste and pollution generation, ecosystem and biodiversity impacts, financing and market barriers, distributional and social justice issues, livelihoods and rights conflicts, social acceptance, and resistance issues, etc. The review has also suggested some possible interventions and solutions to address these trade-offs and challenges and ensure a balanced and holistic approach to RE transition.

This study has analyzed the policy and technical solutions for enhancing the role of RE in advancing sustainable development in South Africa. The study has adopted a literature review approach that synthesizes and evaluates the existing literature on the topic. The study has covered four main aspects: the current status and trends of RE development and consumption; the drivers and barriers for RE adoption and diffusion; the impacts of RE on various dimensions of sustainable development; and the policy and technical solutions for enhancing the role of RE in advancing sustainable development.

The study has found that South Africa has made significant progress in RE development and consumption in recent years, driven by various factors such as cost reduction, policy support, climate action, and energy security. The study has also found that RE can provide multiple benefits for sustainable development, such as reducing greenhouse gas emissions, enhancing energy security, creating jobs, and improving human well-being. However, the study has also found that South Africa still faces some challenges and barriers that limit the potential and impact of RE in advancing sustainable development, such as policy uncertainty, grid constraints, financing gaps, and social resistance.

### 5.4.1   Policy Recommendations

The study has suggested some possible policy and technical solutions to overcome these challenges and barriers and create an enabling environment for RE adoption and diffusion in South Africa. The study has recommended that policymakers should develop or revise national or regional RE policies or strategies that set ambitious but realistic targets for RE share in the energy mix. The study has also recommended that policymakers should align or integrate RE policies with other relevant policies or strategies that address economic development, industrialization, job creation, social justice, climate change mitigation, and adaptation. The study has further recommended that policymakers should engage or involve various stakeholders in policy policy-making and implementation processes, such as government agencies, the private sector, civil society, academia, and local communities. The study has also recommended that policymakers should provide or enhance fiscal and financial incentives and mechanisms to promote RE investment and innovation, such as subsidies, tariffs, taxes, feed-in tariffs, auctions, and quotas. The study has recommended that technical experts should expand or upgrade the grid infrastructure to integrate variable RE sources without compromising stability and quality of supply. The study has also recommended that technical experts should develop or extend transmission and distribution networks to connect RE projects to the grid or off-grid customers. The study has further recommended that technical experts should build or enhance technical skills and expertise for RE planning, installation, operation, and maintenance. The study has also recommended that technical experts should establish or improve standardized quality control and certification systems for RE equipment and services.

## 5.4.2  Future Research

The study has concluded that RE can play a key role in advancing sustainable development in South Africa if the challenges and barriers are addressed and the solutions are implemented effectively. The study has also identified some research gaps and limitations that need to be addressed in future studies. The study has suggested that future research should evaluate and compare the policy and technical solutions for RE in terms of their costs, benefits, risks, trade-offs, and synergies for sustainable development. The study has also suggested that future research should analyze the impacts and outcomes of RE policies and programs in South Africa, especially on the social and environmental dimensions of sustainable development. The study has further suggested that future research should conduct a comprehensive and comparative analysis of the drivers and barriers for different types of RE technologies, such as solar, wind, hydro, biomass, etc., and their suitability and applicability for different contexts and locations in South Africa. The study has also suggested that future research should use participatory and inclusive research methods that involve the perspectives and experiences of various stakeholders, especially the local communities and groups that are affected by or benefit from renewable energy projects. The study has further suggested that future research should adopt interdisciplinary and transdisciplinary research approaches that integrate the natural, social, and engineering sciences to address the complex and interrelated challenges and opportunities of RE transition in South Africa.

However, the review also reveals some research gaps and limitations that need to be addressed in future studies. Some of these gaps include:

- The lack of comprehensive and integrated assessment frameworks and indicators that can capture the multiple and interrelated impacts of RE on sustainable development across different scales, sectors, and contexts.
- The lack of empirical evidence and data on the long-term and cumulative impacts of RE on sustainable development, especially on the environmental and social dimensions, as well as the feedback loops and interactions between different dimensions.
- The lack of comparative analysis of the impacts of different types of RE technologies, such as solar, wind, hydro, biomass, etc., on sustainable development, taking into account their technical, economic, social, and institutional characteristics and implications.
- The lack of participatory and inclusive research methods that involve the perspectives and experiences of various stakeholders, especially the local communities and groups that are affected by or benefit from RE projects, in assessing the impacts of RE on sustainable development.
- The lack of policy analysis and evaluation studies that examine the effectiveness and efficiency of existing policies and programs that support or hinder RE transition about sustainable development goals and targets.

These research gaps indicate the need for more rigorous and comprehensive studies on the impacts of RE on sustainable development that can inform and support policymaking and practice in this field. Such studies can also contribute to the global knowledge base on RE and sustainable development and provide valuable insights and lessons for other countries and regions that are undergoing or planning to undergo similar transitions. The study hopes that its findings and recommendations can contribute to the knowledge base on RE and sustainable development and provide valuable insights and lessons for policymakers, investors, developers, and users of RE technologies in South Africa and beyond.

# References

DMRE (2021) The South African energy sector report. Department of Mineral Resources & Energy. https://www.energy.gov.za/files/media/explained/2021-South-African-Energy-Sector-Report.pdf

Energypedia (2023) South Africa energy situation. Retrieved December 16, 2021 from https://energypedia.info/wiki/South_Africa_Energy_Situation

Handler B, Bazilian M, Hayes M (2021) 5 ways to boost renewable energy investment in developing economies. World Economic Forum. Retrieved from https://www.weforum.org/agenda/2021/06/boost-renewable-energy-investment-in-developing-economies/

Hernandez E, Moreno-Murcia JA, Espín J (2020) Teachers' interpersonal styles and fear of failure from the perspective of physical education students. PLoS ONE 15(6):e0235011

IEA (2019) Renewables 2019: analysis and forecast to 2024. Retrieved from https://www.iea.org/reports/renewables-2019

IRENA (2019) Renewable capacity statistics 2019. Abu Dhabi: International Renewable Energy Agency. https://www.irena.org/publications/2019/Jul/Renewable-energy-statistics-2019

IRENA (2020a) Renewable capacity statistics 2020. Retrieved from https://www.irena.org/public ations/2020/Mar/Renewable-Capacity-Statistics-2020

IRENA. (2020b). Renewable energy prospects for South Africa. Retrieved from https://www. irena.org/-/media/Files/IRENA/Agency/Publication/2020/Nov/IRENA_REmap_South_Afr ica_2020.pdf

IRENA (2020c) Renewable energy market analysis: Africa. International Renewable Energy Agency, Abu Dhabi

IRENA Coalition for Action (2021) Renewable energy: a key driver of low-carbon economic growth in South Africa. Retrieved from https://coalition.irena.org/-/media/Files/IRENA/Coalition-for-Action/Publication/2021/Feb/IRENA_Coalition_for_Action_South_Africa_2021.pdf

Karekezi S, Lata K, Coelho ST (2003) Traditional biomass energy: improving its use and moving to modern energy use. In: Thematic background paper for the international conference for renewable energies. Bonn

Li F, Wang X, Zhang Y, Liang X (2020a) The role of renewable energy consumption and commercial services trade in carbon dioxide reduction: evidence from 25 developing countries. Appl Energy 257:113938

Li F, Cao Y, Wang M, Zhang X, Chen J (2020b) Low-carbon transition of energy systems in developing countries: the case of South Africa. Energy Policy 144:111613

Li F, Liang S, Zhang X, Wang X (2020c) The impact of renewable energy consumption on carbon dioxide emissions: evidence from selected African countries. J Clean Prod 258:120556

Li L et al (2020d) A decision support framework for the design and operation of sustainable urban farming systems. J Clean Prod 268

Murombo T (2022a) Regulatory imperatives for renewable energy: South African perspectives. Journal of African Law 66(1):97–122

Murombo T (2022b) Renewable energy law and policy in South Africa. In: Paterson AR, Kotzé J (eds) Environmental law and governance in Africa: climate change, biodiversity, and human rights. Springer Nature, pp 303–324. https://doi.org/10.1007/978-3-030-46775-7_13

Nhamo G, Nhamo S (2019a) Renewable energy technologies adoption in Africa: a review of trends, policies and challenges across regions. Renew Energy Focus 30:67–79

Nhamo G, Nhamo S (2019b) Renewable energy technologies uptake in South Africa: Policy gaps around human capital development. Sustainability 11(3):906

Nhamo G, Nhamo S (2019c) Renewable energy transitions in South Africa: Policy, institutional and regulatory barriers. In: Nhamo G, Mjimba D (eds) Green economy readiness in South Africa: a focus on the national sphere of government. Springer Nature, pp 97–120

Our World in Data (2023) Renewable energy. Retrieved December 15, 2021, from https://ourwor ldindata.org/renewable-energy

Painuly JP, Park N, Lee M-K, Noh J (2018) Factors affecting renewable energy adoption: a case study of Korea. Renew Sustain Energy Rev 81:3111–3119. https://doi.org/10.1016/j.rser.2017. 06.097

Renewable Energy Agency. https://www.irena.org/media/Files/IRENA/Agency/Publication/2020/Nov/IRENA_RE_market_analysis_Southern_Africa_2020.pdf

WCED (1987) Our common future. Oxford University Press, Oxford

# Chapter 6
# Renewable Energy Dynamics in the North Africa: A System Thinking Approach with the Algerian Case Study

**Abstract** This chapter examines the dynamics of renewable energy (RE) in the North African electricity market, focusing on the Algerian case. The North African countries of Africa, namely Morocco, Algeria, and Tunisia, have a high potential for solar and wind power generation, which could meet their electricity demand and reduce their greenhouse gas emissions. However, they also face various challenges and barriers for RE investment and development, such as policy, technology, economics, behavior, and sustainability. The chapter aims to answer the following questions: What are the policies that support or hinder the development of RE in North Africa? How do they affect the supply and demand of electricity from RE sources? How do they influence regional integration and cooperation in the electricity sector? To answer these questions, the chapter conducts a literature review of articles and journals on this topic and then designs a conceptual model to analyze the dynamics of RE in the Algerian electricity market using system thinking and causal loop diagrams.

*What is in for the readers of this chapter?* This chapter offers the readers the systems thinking approach to analyze the dynamics of renewable energy (RE) in the North African electricity market, focusing on the Algerian case. The readers can learn about the potential and challenges of RE in the region, understand the factors that influence RE growth, apply causal loop diagrams to illustrate the feedback loops and get some best practices and recommendations for promoting and supporting RE development in Algeria. The chapter also discusses the implications and limitations of the analysis, as well as some directions for future research.

## 6.1 Introduction

According to the most widely accepted forecasts, global energy demand will grow moderately by 35% over the next 20 years, mainly due to improvements in energy efficiency. However, this modest increase could still have disastrous environmental impacts, especially for Africa, which is the most vulnerable continent to climate change and desertification. Climate change poses serious threats to Africa's water resources, food security, biodiversity, health, and human security (Boko et al. 2007). Desertification affects about 45% of Africa's land area and threatens the livelihoods of millions of people (UNCCD 2017).

North Africa is a region with diverse energy resources and challenges. On one hand, its energy consumption, especially for electricity, has been growing faster than its supply capacity, creating a gap between demand and supply. According to the International Energy Agency (IEA), North Africa's electricity demand increased by 4.8% per year between 2000 and 2018, while its electricity generation capacity increased by only 3.9% per year (IEA 2020). This gap has resulted in frequent power outages, high electricity prices, and dependence on fossil fuel imports. On the other hand, its energy resources vary from country to country, with some being dependent on energy imports and others being energy exporters. For example, Morocco imports about 90% of its primary energy needs, while Algeria and Libya are major oil and gas producers and exporters (IRENA 2020). This diversity offers opportunities for regional and international cooperation and trade in energy and carbon markets.

North Africa is undergoing a rapid transformation of its electricity markets. The region needs more electricity to support its economic, demographic, and urban growth. However, most of its electricity markets are not liberalized, competitive, or attractive for investors. They are dominated by state monopolies that struggle to fund large-scale electricity projects. These projects are needed to increase the generation capacity and modernize the grids, especially with the abundant RE resources in the region. North Africa has a high potential for solar and wind power generation, which could meet its electricity demand and reduce its greenhouse gas emissions (El-Katiri and Fattouh 2015a, b, c).

The North African countries of Africa, namely Morocco, Algeria, and Tunisia, have recently adopted more market-friendly policies to open up their electricity markets and attract more investment. They have also agreed to integrate and harmonize their national markets with each other. However, the dynamics of RE in this regional market are still unclear. What are the policies that support or hinder the development of RE in North Africa? How do they affect the supply and demand of electricity from renewable sources? How do they influence regional integration and cooperation in the electricity sector? These are some of the questions that this chapter aims to answer. To do so, we will first conduct a literature review of articles and journals on this topic. Then, we will design a conceptual model to analyze the dynamics of RE in the North African electricity market using the Algerian case.

## 6.2  Growth of Renewable Energy—A Brief Overview

The development of RE is influenced by various factors that can act as drivers or barriers in different contexts. The literature on RE has adopted different perspectives and frameworks to analyze these factors and their interactions. Some of the common perspectives are economic, technical, institutional, political, and behavioral (Painuly 2001; Sovacool 2009; Wüstenhagen et al. 2007). Some of the common frameworks are based on clustering the factors into different categories or dimensions, such as organizational, economic, technical, macro, market, social, and governmental (Dragoman 2014a, b; Bellakhal et al. 2016a, b, c; Marques et al. 2010). However, these frameworks are not mutually exclusive or comprehensive, and they may vary depending on the scope, scale, and context of the analysis.

Some studies have also adopted more specific variables to explain the drivers and barriers of RE development in different regions or countries. For example, da Silva et al. (2018) used variables such as $CO_2$ emissions per capita, fossil fuel prices, population growth, GDP per capita, energy use, energy import, and the ratification of the Kyoto Protocol to examine the determinants of RE in Latin America. Similarly, Apergis and Payne (2010) used variables such as real GDP per capita, real gross fixed capital formation per capita, urban population share, trade openness, and financial development to investigate the drivers of RE in 18 OECD countries. Moreover, Ozturk and Acaravci (2013) used variables such as GDP per capita growth rate, urbanization rate, trade openness ratio, and financial development index to explore the barriers to RE in Turkey.

These studies show that there is no single or universal set of factors that can explain the development of RE in all contexts. Rather, the factors are context-specific and dynamic, and they may have different effects depending on the type and level of RE development. Therefore, it is important to adopt a holistic and flexible approach to identify and analyze the drivers and barriers of RE in different situations.

## 6.2.1  A Review of Conceptual Frameworks RE Growth

A critical review of conceptual frameworks linking key variables to the growth of RE is a challenging task, as many different frameworks and variables have been proposed and used in the literature. However, some common themes and issues can be identified and discussed. Here are some possible points for a critical review:

- One of the main issues in developing and applying conceptual frameworks for RE growth is the definition and measurement of RE. RE is a broad term that encompasses various sources, technologies, and applications of energy that are derived from natural resources and have minimal environmental impacts. However, there is no universally agreed definition or classification of RE, and different studies may use different criteria and indicators to measure RE growth (Sovacool et al. 2017). This may create confusion and inconsistency in comparing and evaluating different frameworks and results.
- Another issue is the selection and justification of the key variables that influence RE growth. Many factors can affect RE development, such as economic, technical, institutional, political, social, environmental, and behavioral factors. However, not all factors are equally relevant or important in different contexts and situations.

Therefore, it is essential to identify and explain the rationale for choosing the key variables that are most relevant and significant for the specific research question, scope, scale, and objective of the study. Moreover, it is important to acknowledge the limitations and assumptions of the chosen variables and their interactions (Painuly 2001; Wüstenhagen et al. 2007).

- A third issue is the integration and comparison of different perspectives and frameworks for RE growth. As mentioned earlier, various perspectives and frameworks have been adopted in the literature, such as economic, technical, institutional, political, social, environmental, and behavioral perspectives. Each perspective has its strengths and weaknesses and may emphasize different aspects or dimensions of RE growth. However, none of these perspectives can capture the whole complexity and diversity of RE development. Therefore, it is important to integrate and compare different perspectives and frameworks to gain a more comprehensive and holistic understanding of RE growth. This may require using multiple methods and tools for data collection and analysis, such as literature reviews, surveys, interviews, focus groups, system dynamics modeling, scenario analysis, feasibility assessment, impact evaluation, and policy recommendations (Del Río & Burguillos 2008a, b; da Silva et al. 2018).

### 6.2.2  Example of Conceptual Frameworks Linking Key Variables to the Growth of Renewable

Ata (2015a, b) developed a conceptual model that shows how different factors affect investment decisions in RE. The model shows that renewable policies affect how investors perceive the risk and return of investing in RE, and that technology development affects how much RE technologies cost and how well they perform. The model also implies that the economic approach determines the optimal allocation of resources and the trade-offs between different objectives, such as energy security, environmental protection, and social welfare. The model also considers that investors have different experiences, such as cultural factors, educational backgrounds, and previous exposure to RE investments, that affect how they view and choose RE projects.

The conceptual model of Ata (2015a, b) is consistent with other studies that have identified similar factors as drivers or barriers for RE investment in different contexts. For example, Painuly (2001) classified the barriers to RE penetration into four categories: economic and financial, technical, institutional, and behavioral. Sovacool (2009) argued that the success of RE policies depends on four dimensions: comprehensiveness, consistency, credibility, and communication. Wüstenhagen et al. (2007) introduced the concept of social acceptance of RE innovation, which consists of three aspects: socio-political acceptance, community acceptance, and market acceptance. These studies suggest that RE investment is a complex and multidimensional phenomenon that requires a holistic and flexible approach to address the various challenges and opportunities in different situations.

Del Río and Burguillo (2008a, b) proposed a theoretical framework to analyze the viability and local acceptance of RE projects based on their contributions to local sustainability. They argued that RE projects can have positive impacts on local sustainability in terms of economic growth, environmental sustainability, and social inclusion, by using local resources and creating local development opportunities. However, they also recognized that the benefits of RE projects may not be equally distributed or perceived among different actors and stakeholders, who may have different interests, incentives, and strategies regarding RE development. Therefore, they suggested that stakeholder analysis is a useful tool to understand the dynamics of local acceptance or rejection of RE projects, as well as their implications for the viability and success of the projects. Figure 6.1 illustrates their theoretical framework.

The theoretical framework of Del Río and Burguillo (2008a, b) is consistent with other studies that have emphasized the importance of local sustainability and stakeholder participation in RE development. For example, Walker and Devine-Wright (2008) explored the concept of community renewable energy (CRE), which refers to RE projects that are owned and/or controlled by local communities and that aim to generate social, environmental, and economic benefits for the communities. They identified four dimensions of CRE: ownership, decision-making, participation, and outcomes. They argued that CRE can enhance local sustainability by increasing local control, empowerment, trust, and social capital. Similarly, Rogers et al. (2012) examined the factors that influence community acceptance of wind energy projects in Australia. They found that community acceptance is influenced by perceived impacts on landscape, visual amenities, noise, health, wildlife, property values, tourism, and community cohesion. They also found that community engagement, consultation,

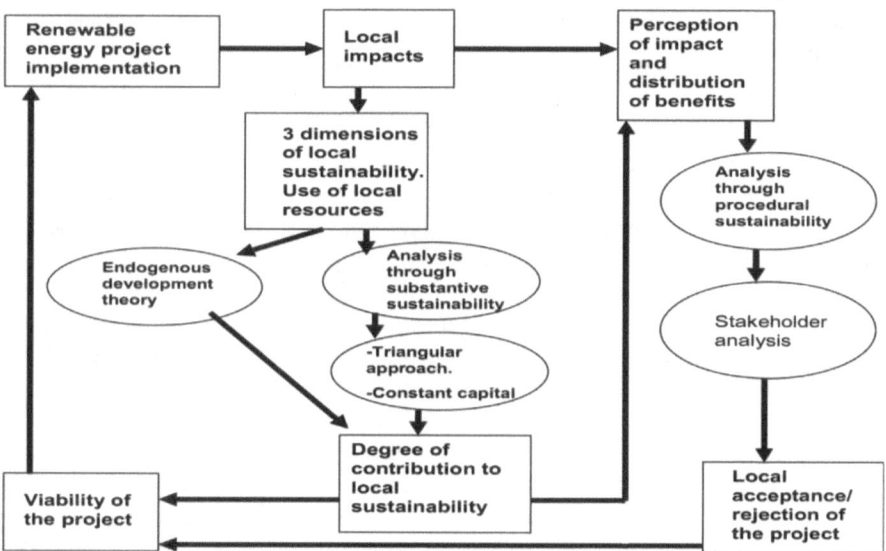

**Fig. 6.1** Theoretical Framework for RE by Del Río and Burguillo (2008a, b)

and participation are crucial for building trust, addressing concerns, and enhancing benefits for the community.

Masini and Menichetti (2012) proposed a conceptual framework to examine the role of behavioral factors in the investment decisions of RE projects. They argued that RE projects are characterized by high uncertainty, long payback periods, and low profitability, which make them less attractive to rational investors. However, they suggested that some investors may have different motivations, preferences, and attitudes toward RE projects, such as environmental awareness, social responsibility, risk aversion, and innovation orientation. These behavioral factors may influence the decision to invest in RE projects, as well as the share of RE in the portfolio. Moreover, they hypothesized that the share of RE in the portfolio may have an impact on the portfolio performance, in terms of risk-adjusted returns and diversification benefits.

The conceptual framework of Masini and Menichetti (2012) is based on three main components: behavioral factors, investment decisions, and portfolio performance. They used a survey method to collect data from a sample of European investors who have invested in RE projects. They used factor analysis to identify the main behavioral factors that affect the investment decision. They also used regression analysis to test the relationship between the behavioral factors and the share of RE in the portfolio. Furthermore, they used portfolio analysis to measure the performance of different portfolios with different shares of RE.

The main findings of Masini and Menichetti (2012) are:

- The behavioral factors that have a significant influence on the decision to invest in RE projects are environmental awareness, social responsibility, risk aversion, and innovation orientation.
- The share of RE in the portfolio is positively related to environmental awareness and social responsibility, and negatively related to risk aversion and innovation orientation.
- The share of RE in the portfolio has a positive impact on the portfolio performance, as it reduces the portfolio risk and increases the portfolio return.

The conceptual framework of Masini and Menichetti (2012) is an original and valuable contribution to the literature on RE investment. It provides a comprehensive and multidimensional perspective on the drivers and outcomes of RE investment. It also offers practical implications for investors, policymakers, and project developers who are interested in promoting and supporting RE development.

Gamel et al. (2017a, b, c) proposed a conceptual framework to examine the factors that influence retail investors' attitudes toward RE investments. They argued that retail investors' attitudes are affected by their evaluation of the regulatory framework, their confidence in politicians and NGOs, their social norms, and their risk aversion. They explained that the regulatory framework is crucial for providing a clear and consistent policy environment that ensures long-term stability and incentives for RE investments (International Energy Agency 2014a, b). They also suggested that politicians and NGOs are important actors in shaping public opinion and the policy agenda on RE issues, as they can act as policy entrepreneurs and advocates for environmental causes (Wüstenhagen et al. 2007; Hrabanski et al. 2013). Moreover, they hypothesized that social norms, such as peer pressure and social responsibility, can influence retail investors' attitudes by creating a sense of moral obligation and collective action towards RE investments. Finally, they assumed that risk aversion, which reflects the degree of uncertainty and variability of returns, can affect retail investors' attitudes by making them more or less willing to invest in RE projects.

Overall, these studies suggest that RE development is not only a technical or economic issue but also a social and political one. Therefore, it is important to adopt a holistic and participatory approach to assess the impacts and implications of RE projects for local sustainability and stakeholder acceptance.

## 6.3 Development of a Conceptual Model for RE Growth in North Africa

We reviewed the relevant literature on RE growth in the North Africa region and identified the following variables that affect RE growth in the region:

- Policy: The regulatory framework provides stability and incentives for RE investment, while politicians and NGOs shape the public opinion and policy agenda on RE issues. A supportive policy environment can increase the perceived level of returns and reduce the perceived level of risk for RE investors, thus increasing their willingness to invest in RE projects. Conversely, a weak or inconsistent policy environment can decrease the perceived level of returns and increase the perceived level of risk for RE investors, thus decreasing their willingness to invest in RE projects. Therefore, the policy has a positive effect on RE investment. Moreover, the growth of RE can create more demand and support for RE policies, as well as more pressure and resistance from competing energy sources.
- Technology: Technology development affects the cost and performance of RE technologies, which can influence the attractiveness and competitiveness of RE projects. A lower cost and higher performance of RE technologies can increase the expected returns and reduce the risk for RE investors, thus increasing their willingness to invest in RE projects. Conversely, a higher cost and lower performance of RE technologies can decrease the expected returns and increase the risk for RE investors, thus decreasing their willingness to invest in RE projects. Therefore, technology has a positive effect on RE investment. Furthermore, the growth of RE can stimulate more innovation and learning in RE technologies, as well as more competition and substitution from other energy technologies.
- Economics: The economic approach determines the optimal allocation of resources and the trade-offs between different objectives, such as energy security, environmental protection, and social welfare. A higher economic value of RE projects can increase the expected returns and reduce the risk for RE investors, thus increasing their willingness to invest in RE projects. Conversely, a lower economic value of RE projects can decrease the expected returns and increase the risk for RE investors, thus decreasing their willingness to invest in RE projects.

Therefore, economics has a positive effect on RE investment. Additionally, the growth of RE can generate more economic benefits and costs for different actors and sectors.

- Behavior: Behavioral factors reflect the motivations, preferences, and attitudes of RE investors, such as environmental awareness, social responsibility, risk aversion, and innovation orientation. These factors can influence the decision to invest in RE projects, as well as the share of RE in the portfolio. A higher level of environmental awareness and social responsibility can increase the willingness to invest in RE projects, while a higher level of risk aversion and innovation orientation can decrease it. Therefore, environmental awareness and social responsibility have a positive effect on RE investment, while risk aversion and innovation orientation have a negative effect on RE investment. Moreover, the growth of RE can affect the behavioral factors of investors by providing more information, experience, and feedback on RE projects.

- Sustainability: Sustainability refers to the impacts and implications of RE projects for local sustainability in terms of economic growth, environmental sustainability, and social inclusion. These impacts can affect the viability and acceptance of RE projects by different stakeholders. A higher level of sustainability can increase the viability and acceptance of RE projects by enhancing their benefits and reducing their costs for local communities. Conversely, a lower level of sustainability can decrease the viability and acceptance of RE projects by increasing their costs and reducing their benefits for local communities. Furthermore, the growth of RE can contribute to or detract from local sustainability by using local resources and creating local development opportunities or challenges.

Based on these identified factors which play a fundamental role in the development and growth of RE in North African countries, we develop a conceptual model for RE growth in North Africa. These factors causally interact with each other to generate the dynamics of RE growth. The following Fig. 6.2 shows the sectorial view of this conceptual model which depicts the interactivity and feedback-orientation among the key variables of RE Growth in North Africa.

Overall, RE is a vital component of the global energy transition and sustainable development. However, the factors that promote or hinder investments in RE sources are not well understood, especially about the governance and institutional quality of different countries (Bellakhal et al., 2016a, b, c). We reviewed some literature on how various authors conceptualized the key variables that drive RE growth, such as policy, technology, economics, behavior, and sustainability. We also examined the RE resources and potential in the North Africa region, which has a high solar and wind energy advantage but faces many challenges and opportunities for RE development. Based on our literature review and regional analysis, we proposed a conceptual framework for the North African region that links the main factors affecting RE growth. In the next section, we will test this conceptual framework with data on the variables used from a North African country, Algeria.

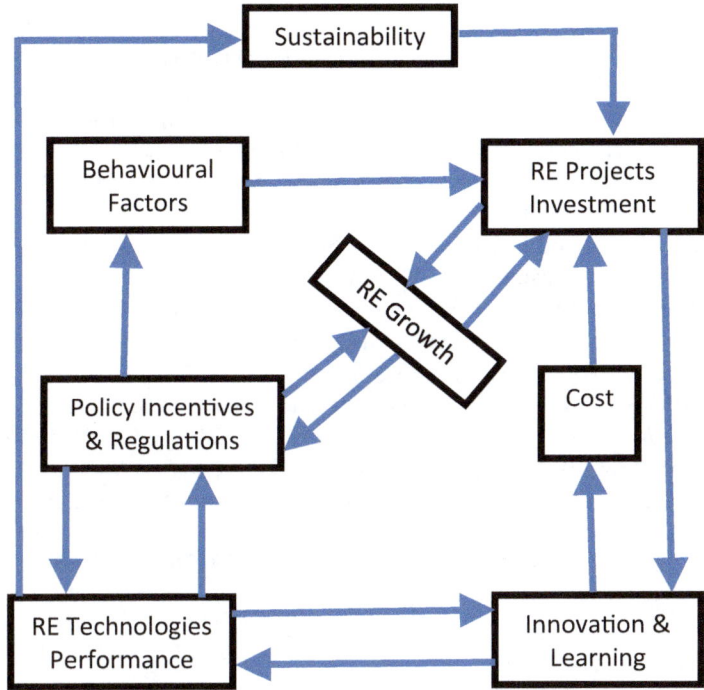

**Fig. 6.2**  A conceptual model for RE growth in North Africa

## 6.4  Analyzing the Case of RE Growth in Algeria

Now, we can apply this conceptual model to analyze the supply, demand, and growth of RE in Algeria. This model can help us understand the main factors that affect RE investment and development in Algeria, as well as the feedback effects of RE growth on these factors. The model can also help us compare Algeria with other North African countries in terms of their RE potential and performance.

To apply this model, we need to collect data on the variables used in the model for Algeria and other North African countries. We can use various sources of data, such as official statistics, reports, surveys, and academic papers.

### 6.4.1  Dynamics of RE Demand and Supply in Algeria

The current and projected levels of RE supply, demand, and growth in Algeria and other North African countries vary depending on the sources and methods of estimation. According to the International Renewable Energy Agency (IRENA 2021a), the total RE supply in Algeria was 1.6 million tonnes of oil equivalent (Mtoe) in 2019, accounting for 2.3% of the total primary energy supply (TPES) of 69.6 Mtoe. The main sources of RE supply were solar (0.8 Mtoe), bioenergy (0.5 Mtoe), and wind (0.3 Mtoe), as is shown in Fig. 6.3. The total RE supply in North Africa was 8.9 Mtoe in 2019, accounting for 3.4% of the TPES of 261.6 Mtoe. The main sources of RE supply were bioenergy (4.7 Mtoe), hydro (2.1 Mtoe), wind (1.2 Mtoe), and solar (0.9 Mtoe). The leading countries in RE supply were Morocco (3.5 Mtoe), Egypt (2.8 Mtoe), and Algeria (1.6 Mtoe).

The total RE demand in Algeria was 1.5 Mtoe in 2019, accounting for 2.4% of the total final energy consumption (TFEC) of 62.7 Mtoe (IRENA 2021b). Figure 6.4 shows that the main sectors of RE demand were electricity generation (0.8 Mtoe), industry (0.4 Mtoe), and transport (0.2 Mtoe). The total RE demand in North Africa was 8.5 Mtoe in 2019, accounting for 3.6% of the TFEC of 237.4 Mtoe. The main sectors of RE demand were electricity generation (4.7 Mtoe), industry (1.7 Mtoe), transport (1.2 Mtoe), and residential (0.7 Mtoe). The leading countries in RE demand were Morocco (3.4 Mtoe), Egypt (2.7 Mtoe), and Algeria (1.5 Mtoe).

The total RE growth in Algeria was 11% in 2019, compared to the average annual growth rate of 5% over the period 2010–2019. The main drivers of RE growth were solar (+24%), wind (+14%), and bioenergy (+6%). The total RE growth in North

**Fig. 6.3** Sources of RE supply in Algeria 2019

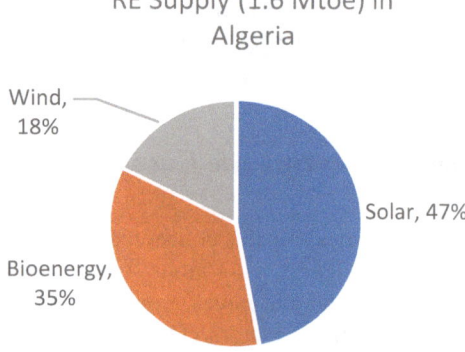

**Fig. 6.4** Sector-wide RE Supply in Algeria in 2019

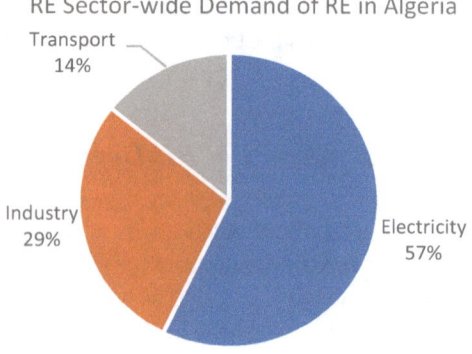

Africa was 10% in 2019, compared to the average annual growth rate of 7% over the period 2010–2019. The main drivers of RE growth were wind (+17%), solar (+16%), hydro (+10%), and bioenergy (+5%). The leading countries in RE growth were Morocco (+14%), Egypt (+11%), and Algeria (+11%).

The projected levels of RE supply, demand, and growth in Algeria and other North African countries depend on the scenarios and assumptions used for the future development of the energy sector. According to IRENA's RE Outlook for Africa, under the current policies and plans scenario, the total RE supply in Algeria could reach 3.8 Mtoe by 2030, accounting for 5% of the TPES of 76 Mtoe, while under the accelerated case scenario, it could reach 10.4 Mtoe by 2030, accounting for 13% of the TPES of 80 Mtoe.

Similarly, under the current policies and plans scenario, the total RE supply in North Africa could reach 20.6 Mtoe by 2030, accounting for 7% of the TPES of 292 Mtoe, while under the accelerated case scenario, it could reach 49.8 Mtoe by 2030, accounting for 15% of the TPES of 333 Mtoe. The main sources of RE supply under both scenarios would be solar, wind, hydro, and bioenergy, with varying shares depending on the country and the level of ambition.

The projected levels of RE demand and growth in Algeria and other North African countries would follow similar trends as those of RE supply, depending on the scenarios and assumptions used for the future development of the energy sector. Under both scenarios, the main sectors of RE demand would be electricity generation, industry, transport, and residential, with varying shares depending on the country and the level of ambition.

## 6.4.2  Using Systems Thinking to Understand the Dynamics of RE Growth in Algeria

The CLD in Fig. 6.5 shows four feedback loops, B1, B2, B3, and R1 that affect RE growth in Algeria. For example, there is a balancing feedback loop, B1, between fossil fuel price and fossil fuel consumption. An increase in fossil fuel prices reduces fossil fuel consumption, which lowers the demand and price of fossil fuels.

The second balancing feedback loop, B2, is between fossil fuel consumption and greenhouse gas emissions. An increase in fossil fuel consumption increases greenhouse gas emissions, which contributes to climate change and environmental degradation, which creates more pressure and incentives to reduce fossil fuel consumption.

The third balancing feedback loop between fossil fuel consumption and energy security. An increase in fossil fuel consumption reduces energy security, as it increases the dependence on imported energy sources and exposes the country to external shocks and uncertainties, which creates more pressure and incentives to reduce fossil fuel consumption.

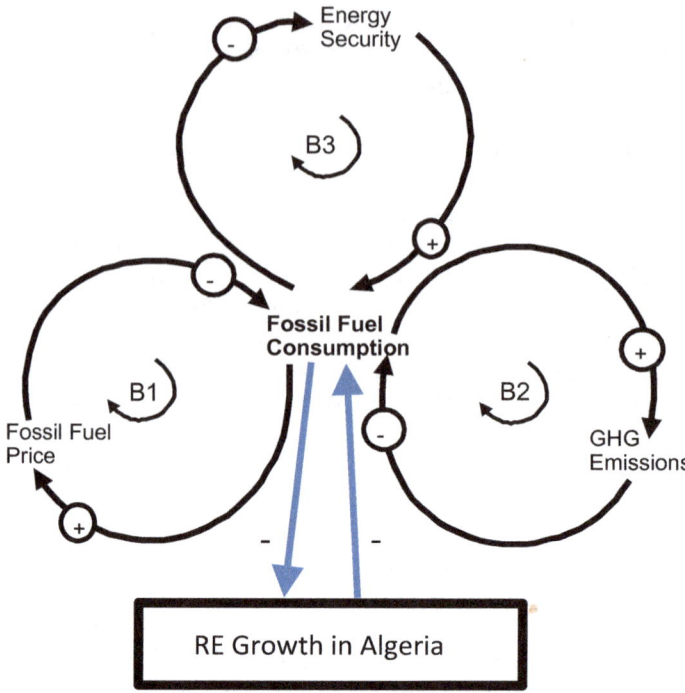

**Fig. 6.5** Dynamics of Traditional Fuel and RE Growth in Algeria

Finally, a reinforcing feedback loop, R1, exists between fossil fuel consumption and RE growth in Algeria: with more fossil fuel consumption, the share of RE generation declines and vice versa.

In Fig. 6.6, a CLD describes three feedback loops, R1, 2, and R3, which are responsible for RE growth in Algeria via RE-supporting energy policies. In R1, a reinforcing loop between RE policy and RE investment, a supportive RE policy increases RE investment, which increases the supply and competitiveness of RE projects, which creates more demand and support for RE policy.

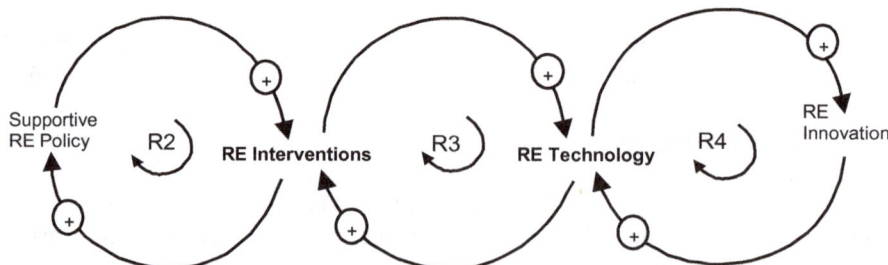

**Fig. 6.6** Dynamics of RE Policy-RE Investment-RE Technology in Algeria

The second feedback loop, R2 is a reinforcing loop between RE investment and RE technology. An increase in RE investment enhances the development and diffusion of RE technology, which lowers the cost and improves the performance of RE projects, which increases the attractiveness and profitability of RE investment.

Finally, the third feedback loop, R3 is a reinforcing loop between RE technology and RE innovation. An increase in RE technology fosters more innovation and learning in RE, which leads to discoveries and breakthroughs in RE, which enhances the cost and performance of RE technology.

These feedback loops, in Figs. 6.5 and 6.6 can help us understand how different factors interact and influence each other in the system of RE growth in Algeria. They can also help us identify potential leverage points or interventions that can improve or optimize the energy system's behavior and outcomes. For example:

- To reduce fossil fuel consumption and greenhouse gas emissions in Algeria, policymakers can intervene by increasing the fossil fuel price through taxes or subsidies removal, or by increasing the awareness and education of consumers about the environmental impacts of fossil fuels.
- To increase energy security and reduce dependence on imported energy sources in Algeria, the decision makers can intervene by diversifying the energy mix and increasing the share of domestic RE sources, or by enhancing regional cooperation and integration with other countries on energy issues.
- To promote and support RE development in Algeria, we the decision makers can intervene by implementing supportive policies and regulations that provide incentives and stability for RE investors and developers, or by investing more in research and development activities and capacities that foster innovation and learning in RE technologies.

## 6.5  The Drivers and Barriers for RE Investment and Development in Algeria

The main drivers and barriers for RE investment and development in Algeria and other North African countries are related to the factors that we identified in our conceptual model (please see Fig. 6.2), such as policy, technology, economics, behavior, and sustainability. However, the relative importance and impact of these factors may vary across countries and over time, depending on the specific context and conditions of each country. Here, we summarize some of the recent literature that has examined the drivers and barriers for RE investment and development in Algeria and other North African countries.

- *Energy policy*: The energy policy environment is a crucial factor in creating a stable and supportive framework for RE investment and development. A clear and consistent policy vision, strategy, and plan can provide long-term certainty and incentives for RE investors and developers, as well as reduce the perceived level of risk and uncertainty. Moreover, a coordinated and integrated policy approach can address the multiple dimensions and objectives of RE development, such as energy security, environmental protection, social welfare, and regional integration. However, the policy environment in Algeria and other North African countries is often characterized by weak or inconsistent policies, lack of coordination and integration, political instability and conflicts, institutional inefficiencies and corruption, and regulatory barriers and bottlenecks (Bellakhal et al. 2016a, b, c; Benkhedda et al. 2020; El-Katiri and Fattouh 2015a, b, c; Gamel et al. 2017a, b, c; Sghaier et al. 2021).

- *Technology*: Technology development is another key factor for enhancing the cost and performance of RE technologies, which can influence the attractiveness and competitiveness of RE projects. A lower cost and higher performance of RE technologies can increase the expected returns and reduce the risk for RE investors and developers, as well as increase the affordability and accessibility of RE for consumers. Furthermore, technology development can foster innovation and learn in RE technologies, as well as facilitate technology transfer and diffusion across countries. However, the technology development in Algeria and other North African countries is often constrained by limited research and development capacities and activities, lack of skilled human resources and technical expertise, high dependence on foreign technologies and suppliers, low quality of infrastructure and services, and insufficient adaptation to local conditions and needs (Bellakhal et al. 2016a, b, c; Benkhedda et al. 2020; El-Katiri and Fattouh 2015a, b, c; Gamel et al. 2017a, b, c; Sghaier et al. 2021).

- *Economics*: The economic approach is another important factor for determining the optimal allocation of resources and the trade-offs between different objectives, such as energy security, environmental protection, and social welfare. A higher economic value of RE projects can increase the expected returns and reduce the risk for RE investors and developers, as well as increase the economic benefits and opportunities for different actors and sectors. Moreover, an economic approach can help evaluate the costs and benefits of RE projects from a social perspective, taking into account the externalities and spillovers of RE development. However, the economic approach in Algeria and other North African countries is often challenged by low or volatile prices of fossil fuels, high subsidies for conventional energy sources, high costs of financing and capital for RE projects, low level of income and purchasing power for consumers, lack of market mechanisms and instruments for RE promotion, and weak integration with regional markets (Bellakhal et al. 2016a, b, c; Benkhedda et al. 2020; El-Katiri and Fattouh 2015a, b, c; Gamel et al. 2017a, b, c; Sghaier et al. 2021).

- *Behavior*: The behavioral factors are another relevant factor for reflecting the motivations, preferences, and attitudes of RE investors, developers, consumers, and other stakeholders. These factors can influence the decision to invest in or adopt RE projects, as well as the share of RE in the energy mix. A higher level of environmental awareness and social responsibility can increase the willingness to invest in or adopt RE projects, while a higher level of risk aversion and innovation orientation can decrease it. Moreover, behavioral factors can be affected by the information, experience, and feedback on RE projects, which can shape the perceptions and expectations of different stakeholders. However, the behavioral factors in Algeria and other North African countries are often influenced by low level of awareness and knowledge of RE benefits and opportunities, high level of resistance and inertia to change the established energy habits and practices, low level of trust and confidence in RE technologies and providers, and low level of participation and involvement in RE development (Bellakhal et al. 2016a, b, c; Benkhedda et al. 2020; El-Katiri and Fattouh 2015a, b, c; Gamel et al. 2017a, b, c; Sghaier et al. 2021).

- *Sustainability*: The sustainability factor is another significant factor for assessing the impacts and implications of RE projects for local sustainability in terms of economic growth, environmental sustainability, and social inclusion. These impacts can affect the viability and acceptance of RE projects by different stakeholders. A higher level of sustainability can increase the viability and acceptance of RE projects by enhancing their benefits and reducing their costs for local communities. Conversely, a lower level of sustainability can decrease the viability and acceptance of RE projects by increasing their costs and reducing their benefits for local communities. Furthermore, the growth of RE can contribute to or detract from local sustainability by using local resources and creating local development opportunities or challenges. However, the sustainability factor in Algeria and other North African countries is often overlooked or underestimated by the decision makers and the public, who tend to focus more on the short-term economic gains or losses of RE projects, rather than the long-term environmental and social impacts and benefits of RE development (Bellakhal et al. 2016a, b, c; Benkhedda et al. 2020; El-Katiri and Fattouh 2015a, þ, c; Gamel et al. 2017a, b, c; Sghaier et al. 2021).

In summary, the main drivers and barriers for RE investment and development in Algeria and other North African countries are multifaceted and interrelated, and they require a holistic and flexible approach to address them effectively and efficiently. The conceptual model that we proposed can help us understand and analyze these factors and their interactions, as well as their implications for RE growth in the region.

### 6.5.1   Using Systems Thinking to Analyze Drivers and Barriers of RE Growth in Algeria

The CLD in Fig. 6.7 shows five feedback loops, R5, B4, R6, B5, and R7 that represent drivers and barriers of RE growth in Algeria. In R5, a positive reinforcing loop between energy policy and RE investment and development: A clear and consistent energy policy can provide long-term certainty and incentives for RE investors and developers, which can increase the level of RE investment and development in the country. This, in turn, can create more demand and support for RE policies, which can

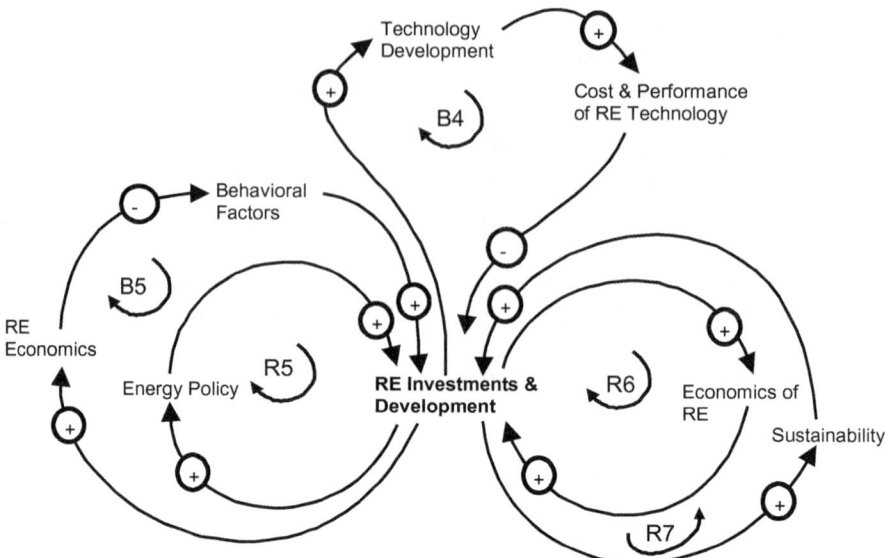

**Fig. 6.7** A CLD of drivers and barriers of RE growth in Algeria

further enhance the energy policy environment for RE investment and development. This loop can be labeled as R5: Policy support for RE growth.

The first negative balancing loop, B4, between technology development and RE investment and development: A lower cost and higher performance of RE technologies can increase the attractiveness and competitiveness of RE projects, which can increase the level of RE investment and development in the country. However, this can also reduce the need and motivation for further technology development, as the existing technologies may be sufficient or dominant in the market. This can lower the level of technology development in the country, which can eventually affect the cost and performance of RE technologies. This loop can be labeled as B4: Technology improvement for RE growth.

In another positive reinforcing loop, R6, between economics and RE investment and development, a higher economic value of RE projects can increase the expected returns and reduce the risk for RE investors and developers, which can increase the level of RE investment and development in the country. This, in turn, can create more economic benefits and opportunities for different actors and sectors, such as job creation, income generation, tax revenue, etc. This can increase the economic value of RE projects, which can further enhance the attractiveness and competitiveness of RE projects. This loop can be labeled as R6: Economic value for RE growth.

There is another negative balancing loop, B5, between behavior and RE investment and development. In this feedback loop, a higher level of environmental awareness and social responsibility can increase the willingness to invest in or adopt RE projects, which can increase the level of RE investment and development in the country. However, this can also increase the expectations and demands for RE projects, such as quality, reliability, affordability, etc. This can affect the satisfaction and acceptance of RE projects by different stakeholders, which can influence their behavior towards RE projects. This loop can be labeled as B2: Behavioral factors for RE growth.

Finally, in Fig. 6.6, the CLD represents a positive reinforcing loop, R7, between sustainability and RE investment and development where a higher level of sustainability of RE projects can increase their viability and acceptance by different stakeholders, which can increase the level of RE investment and development in the country. This, in turn, can contribute to local sustainability in terms of economic growth, environmental protection, and social inclusion. This can increase the sustainability of RE projects, which can further enhance their viability and acceptance by different stakeholders. This loop can be labeled as R3: Sustainability for RE growth.

This section has provided a comprehensive review of the main drivers and barriers for RE investment and development in Algeria and other North African countries, based on the existing literature and our conceptual model. We have identified five factors that can influence RE growth in the region: energy policy, technology, economics, behavior, and sustainability. These factors are interrelated and dynamic, and they can have positive or negative effects on RE growth, depending on the specific context and conditions of each country. To better understand and analyze these factors and their interactions, we have used a system thinking approach, which is a holistic and flexible method that can capture the complexity and diversity of RE systems. We have used causal loop diagrams (CLDs) to illustrate the feedback loops that can enhance or hinder RE growth in the region. These CLDs show how the changes in one factor can affect the changes in another factor, and how these changes can reinforce or balance each other over time. By using CLDs, we can identify the key leverage points and intervention strategies that can improve the performance and outcomes of RE systems in the region.

## 6.6   How Does RE Growth Affect the Policy, Technology, Economics, Behavior, and Sustainability Factors in Algeria

RE growth can affect the policy, technology, economics, behavior, and sustainability factors in Algeria in various ways, depending on the direction and magnitude of the change. Here, we summarize some of the possible effects based on the literature:

- *Policy*: RE growth can create more demand and support for RE policies, as well as more pressure and resistance from competing energy sources. For example, RE growth can increase public awareness and acceptance of RE benefits and opportunities, as well as the political will and commitment to implement RE policies. However, RE growth can also increase the conflicts and trade-offs between different policy objectives, such as energy security, environmental protection, and social welfare. Moreover, RE growth can challenge the existing institutional and regulatory frameworks, which may not be adequate or flexible enough to accommodate the rapid and diverse changes in the energy sector (Bellakhal et al. 2016a, b, c; Benkhedda et al. 2020; El-Katiri and Fattouh 2015a, b, c).

- **Technology**: RE growth can stimulate more innovation and learning in RE technologies, as well as more competition and substitution from other energy technologies. For example, RE growth can increase the research and development activities and capacities in RE technologies, as well as the technology transfer and diffusion across countries. However, RE growth can also increase the technological uncertainties and risks, as well as the technological gaps and dependencies. Furthermore, RE growth can require more adaptation and integration of RE technologies to the local conditions and needs, as well as to the existing infrastructure and services (Bellakhal et al. 2016a, b, c; Benkhedda et al. 2020; El-Katiri and Fattouh 2015a, b, c).

- **Economics**: RE growth can generate more economic benefits and costs for different actors and sectors. For example, RE growth can increase the economic value and competitiveness of RE projects, as well as the economic opportunities and diversification for different actors and sectors. However, RE growth can also increase the economic uncertainties and risks, as well as the economic trade-offs and conflicts. Moreover, RE growth can affect the prices and markets of energy sources, as well as the allocation and distribution of resources (Bellakhal et al. 2016a, b, c; Benkhedda et al. 2020; El-Katiri and Fattouh 2015a, b, c).

- **Behavior**: RE growth can affect the behavioral factors of investors, developers, consumers, and other stakeholders by providing more information, experience, and feedback on RE projects. For example, RE growth can increase the environmental awareness and social responsibility of different stakeholders, as well as their willingness to invest in or adopt RE projects. However, RE growth can also increase the resistance and inertia to change the established energy habits and practices, as well as the trust and confidence issues in RE technologies and providers. Furthermore, RE growth can influence the preferences and attitudes of different stakeholders towards RE projects (Bellakhal et al. 2016a, b, c; Benkhedda et al. 2020; El-Katiri and Fattouh 2015a, b, c; Gamel et al. 2017a, b, c).

- **Sustainability**: RE growth can contribute to or detract from local sustainability by using local resources and creating local development opportunities or challenges. For example, RE growth can enhance local sustainability by reducing greenhouse gas emissions and environmental pollution, increasing energy security and access, creating jobs and income, and improving social inclusion and cohesion.

However, RE growth can also detract from local sustainability by causing land use and water conflicts, displacing local communities and livelihoods, creating waste and health problems, and increasing social inequalities and tensions. Moreover, RE growth can have spillover effects on regional and global sustainability (Bellakhal et al. 2016a, b, c; Benkhedda et al. 2020; El-Katiri and Fattouh 2015a, b, c; Gamel et al. 2017a, b, c).

Overall, RE growth can have complex and dynamic effects on the policy, technology, economics, behavior, and sustainability factors in Algeria. These effects can be positive or negative, depending on the context and conditions of each factor. The conceptual model that we proposed can help us understand and analyze these effects and their interactions, as well as their implications for RE development in Algeria.

### 6.6.1   Systems Thinking and Factors Affecting RE Growth in Algeria

The CLD in Fig. 6.8 shows five feedback loops that affect RE growth in Algeria: three balancing and two reinforcing. The first balancing loop, B6, is between RE growth and economics. In this feedback loop, an increase in RE growth generates more economic benefits and costs for different actors and sectors, which affects the prices and markets of energy sources, as well as the allocation and distribution of resources. This can create economic uncertainties and risks, as well as trade-offs and conflicts, which can reduce the attractiveness and profitability of RE projects.

The second balancing loop, B7, is between RE growth and behavior where an increase in RE growth affects the behavioral factors of investors, developers, consumers, and other stakeholders by providing more information, experience, and feedback on RE projects. This can increase the environmental awareness and social responsibility of different stakeholders, as well as their willingness to invest in or adopt RE projects. However, this can also increase the resistance and inertia to change the established energy habits and practices, as well as the trust and confidence issues

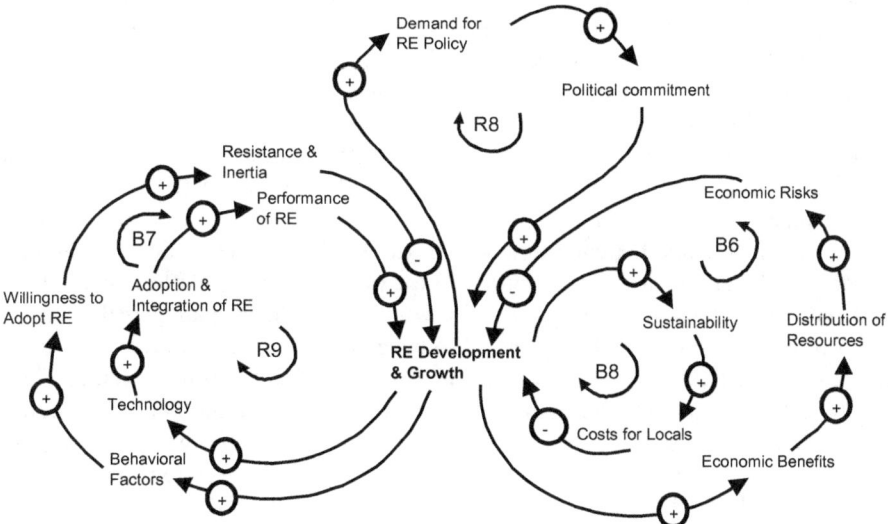

**Fig. 6.8**  A CLD of factors affecting RE growth in Algeria

in RE technologies and providers. This can reduce the share and Acceptance of RE projects.

The third balancing loop, B8, is between RE growth and sustainability. Here, an increase in RE growth contributes to or detracts from local sustainability by using local resources and creating local development opportunities or challenges. This can affect the viability and acceptance of RE projects by different stakeholders by enhancing or reducing their benefits and costs for local communities. This can also have spillover effects on regional and global sustainability by reducing or increasing greenhouse gas emissions and environmental pollution.

The first growth propelling, a reinforcing loop, R8 is between RE growth and policy. An increase in RE growth creates more demand and support for RE policies, as well as more pressure and resistance from competing energy sources. This can increase public awareness and acceptance of RE benefits and opportunities, as well as the political will and commitment to implement RE policies. This can also increase the conflicts and trade-offs between different policy objectives, such as energy security, environmental protection, and social welfare. This can challenge the existing institutional and regulatory frameworks, which may not be adequate or flexible enough to accommodate the rapid and diverse changes in the energy sector. This can create more demand and support for policy improvement or reform.

Finally, a reinforcing loop, R9, is between RE growth and technology. In this positive feedback loop, an increase in RE growth stimulates more innovation and learning in RE technologies, as well as more competition and substitution from other energy technologies. This can increase the research and development activities and capacities in RE technologies, as well as the technology transfer and diffusion across countries. This can also increase the technological uncertainties and risks, as well as the technological gaps and dependencies. This can require more adaptation and integration of RE technologies to the local conditions and needs, as well as to the existing infrastructure and services. This can enhance the cost and performance of RE technologies.

In conclusion, this section has used a system thinking approach to analyze the factors affecting RE growth in Algeria, using a causal loop diagram (CLD) to illustrate the feedback loops that can enhance or hinder RE growth in the country. We have identified five factors that can influence RE growth in Algeria: policy, technology, economics, behavior, and sustainability. These factors are interrelated and dynamic, and they can have positive or negative effects on RE growth, depending on the specific context and conditions of the country. We have also identified five feedback loops that can affect RE growth in Algeria: three balancing and two reinforcing. These feedback loops show how the changes in one factor can affect the changes in another factor, and how these changes can reinforce or balance each other over time. By using CLDs, we can identify the key leverage points and intervention strategies that can improve the performance and outcomes of RE systems in the country. The next section will discuss the best practices and recommendations for RE development and growth in Algeria

## 6.7  Best Practices and Recommendations for RE Development in Algeria

The best practices and recommendations for promoting and supporting RE development in Algeria are based on the literature that has analyzed the drivers and barriers for RE investment and development in Algeria and other North African countries. Some of the main recommendations are:

- *Policy*: Algeria should adopt a clear and consistent policy vision, strategy, and plan for RE development, that provides long-term certainty and incentives for RE investors and developers, as well as reduces the perceived level of risk and uncertainty. Algeria should also coordinate and integrate its RE policy with other policy objectives, such as energy security, environmental protection, and social welfare, and address the potential conflicts and trade-offs between them. Moreover, Algeria should improve its institutional and regulatory frameworks, to enhance the efficiency and transparency of the energy sector, and remove the barriers and bottlenecks for RE projects (Bellakhal et al. 2016a, b, c; Benkhedda et al. 2020; El-Katiri and Fattouh 2015a, b, c).
- *Technology*: Algeria should invest more in research and development activities and capacities in RE technologies, to enhance the cost and performance of RE technologies, as well as foster innovation and learning. Algeria should also develop its skilled human resources and technical expertise in RE technologies, to reduce its dependence on foreign technologies and suppliers. Furthermore, Algeria should adapt and integrate its RE technologies to the local conditions and needs, as well as to the existing infrastructure and services (Bellakhal et al. 2016a, b, c; Benkhedda et al. 2020; El-Katiri and Fattouh 2015a, b, c).
- *Economics*: Algeria should diversify its energy sources and reduce its reliance on fossil fuels, to increase its energy security and resilience to price volatility. Algeria should also phase out its subsidies for conventional energy sources, to increase the competitiveness and attractiveness of RE projects. Moreover, Algeria should facilitate access to financing and capital for RE projects, to lower the costs of investment. Furthermore, Algeria should implement market mechanisms and instruments for RE promotion, such as feed-in tariffs, net metering, and renewable portfolio standards. Additionally, Algeria should enhance its integration with regional markets, to increase it is export opportunities and cooperation with other countries (Bellakhal et al. 2016a, b, c; Benkhedda et al. 2020; El-Katiri and Fattouh 2015a, b, c).
- *Behavior*: Algeria should raise awareness and knowledge of RE benefits and opportunities among different stakeholders, such as investors, developers, consumers, and policymakers. Algeria should also encourage the participation and involvement of different stakeholders in RE development, such as through public consultation, community engagement, and stakeholder dialogue. Moreover, Algeria should build trust and confidence in RE technologies and providers, such as through quality standards, certification schemes, and consumer protection measures. Furthermore, Algeria should provide information, experience, and

feedback on RE projects, such as through monitoring, evaluation, and dissemination activities (Bellakhal et al. 2016a, b, c; Benkhedda et al. 2020; El-Katiri and Fattouh 2015a, b, c; Gamel et al. 2017a, b, c).

- *Sustainability*: Algeria should assess the impacts and implications of RE projects for local sustainability in terms of economic growth, environmental sustainability, and social inclusion. Algeria should also balance the benefits and costs of RE projects for different stakeholders, and ensure a fair and equitable distribution of resources and opportunities. Moreover, Algeria should mitigate the negative effects of RE projects on local communities and livelihoods, such as land use and water conflicts, displacement, waste, and health problems. Furthermore, Algeria should contribute to regional and global sustainability by reducing greenhouse gas emissions and environmental pollution, and supporting international cooperation and commitments on climate change (Bellakhal et al. 2016a, b, c; Benkhedda et al. 2020; El-Katiri and Fattouh 2015a, b, c; Gamel et al. 2017a, b, c).

In conclusion, this section has provided some best practices and recommendations for promoting and supporting RE development in Algeria, based on the literature that has analyzed the drivers and barriers to RE investment and development in Algeria and other North African countries. We have suggested some policy, technology, economics, behavior, and sustainability measures that can enhance the attractiveness and competitiveness of RE projects in Algeria, as well as increase the benefits and opportunities for different actors and sectors. We have also highlighted the importance of regional integration and cooperation in the electricity sector, as well as the contribution of RE development to local, regional, and global sustainability. These recommendations are not exhaustive or definitive, but rather indicative and suggestive, as they may need to be adapted and modified according to the specific context and conditions of each country. The next section will summarize the main findings and conclusions of this chapter, as well as discuss some limitations and challenges of our analysis and some directions for future research.

## 6.8  Summary, Limitations, and Future Research

This chapter has examined the dynamics of RE in the Maghreb (i.e., North Africa) electricity market, focusing on the Algerian case. The Maghreb countries of Africa, namely Morocco, Algeria, and Tunisia, have a high potential for solar and wind power generation, which could meet their electricity demand and reduce their greenhouse gas emissions. However, they also face various challenges and barriers for RE investment and development, such as policy, technology, economics, behavior, and sustainability. The chapter aimed to answer the following questions: What are the policies that support or hinder the development of RE in North Africa? How do they affect the supply and demand of electricity from RE sources? How do they influence regional integration and cooperation in the electricity sector? To answer these questions, the chapter conducted a literature review of articles and journals on this topic and then designed a conceptual model to analyze the dynamics of RE in the Algerian electricity market using system thinking and causal loop diagrams. The chapter also provided some best practices and recommendations for promoting and supporting RE development in Algeria.

The main findings and conclusions of this chapter are:

- The main drivers and barriers for RE investment and development in Algeria and other North African countries are multifaceted and interrelated, and they require a holistic and flexible approach to address them effectively and efficiently. The conceptual model that we proposed can help us understand and analyze these factors and their interactions, as well as their implications for RE growth in the region.
- The dynamics of RE in the Maghreb electricity market are complex and diverse, and they depend on the specific context and conditions of each country. The causal loop diagrams that we developed can illustrate the feedback loops that can enhance or hinder RE growth in the region. These feedback loops show how the changes in one factor can affect the changes in another factor, and how these changes can reinforce or balance each other over time.
- The best practices and recommendations for promoting and supporting RE development in Algeria are based on the literature that has analyzed the drivers and barriers for RE investment and development in Algeria and other North African countries. We have suggested some policy, technology, economics, behavior, and sustainability measures that can enhance the attractiveness and competitiveness of RE projects in Algeria, as well as increase the benefits and opportunities for different actors and sectors. We have also highlighted the importance of regional integration and cooperation in the electricity sector, as well as the contribution of RE development to local, regional, and global sustainability.

However, this chapter also has some limitations and challenges that need to be acknowledged and addressed. Some of these limitations are:

- The literature review that we conducted was not exhaustive or systematic, but rather indicative or selective. We focused on some of the most relevant articles and journals that we could find on this topic, but we may have missed some important sources or perspectives that could enrich our analysis.
- The conceptual model that we designed was not empirical or quantitative, but rather theoretical or qualitative. We did not use any data or evidence to support our assumptions or hypotheses, but we relied on our knowledge and understanding of the topic. Therefore, our model may not be accurate or realistic enough to capture the actual dynamics of RE in the Maghreb electricity market.
- The causal loop diagrams that we developed were not validated or tested, but rather exploratory or illustrative. We did not use any methods or tools to check the validity or reliability of our diagrams, but we used our judgment and intuition to draw them. Therefore, our diagrams may not be consistent or comprehensive enough to represent the feedback loops that affect RE growth in the region.

- The best practices and recommendations that we provided were not definitive or prescriptive, but rather indicative or suggestive. We did not consider all the possible scenarios or outcomes that could result from implementing our recommendations, but we based them on our analysis and evaluation of the situation. Therefore, our recommendations may not be feasible or effective enough to address the challenges and barriers to RE development in Algeria.

Some directions for future research that can overcome these limitations are:

- Conducting a more systematic and comprehensive literature review on this topic, using different sources and methods to collect and analyze relevant information.
- Developing a more empirical and quantitative conceptual model on this topic, using data and evidence to support our assumptions or hypotheses.
- Validating or testing our causal loop diagrams on this topic, using methods or tools to check their validity or reliability.
- Considering different scenarios or outcomes that could result from implementing our recommendations on this topic, using methods or tools to evaluate their feasibility or effectiveness.

We hope that this chapter has contributed to advancing the knowledge and understanding of RE dynamics in the North African electricity market, especially in Algeria. We also hope that this chapter has provided some useful insights and guidance for policymakers and practitioners who are interested in promoting RE development in Algeria. We believe that RE development is not only a technical or economic issue but also a social or environmental one. Therefore, we need to adopt a holistic and flexible approach that can address the multiple dimensions and objectives of RE development, as well as the complexity and diversity of RE systems.

# References

African Development Bank (2012) African development report 2012: Towards green growth in Africa. Abidjan: African Development Bank. https://am.afdb.org/en/past-annual-meetings/2013-annual-meetings/programme/africa-development-report-2012

Apergis N, Payne JE (2010) Renewable energy consumption and economic growth: evidence from a panel of OECD countries. Energy Policy 38(1):656–660

Ata NK (2015a) The impact of government policies in the renewable energy investment: developing a conceptual framework and qualitative analysis. Glob Adv Res J Manag Bus Stud 4(2):067–081

Ata R (2015b) Renewable energy policy and investment in Turkey: a case study in the context of the European Union. Renew Sustain Energy Rev 52:300–310

Bellakhal R, Ben Aissa MS, Goaied M (2016a) Renewable energy adoption and economic growth: evidence from MENA countries. Int J Green Energy 13(15):1565–1573

Bellakhal R, Ben Youssef A, M'nasri A (2016b) Renewable energy, environmental policy and growth in MENA countries. Int J Sustain Dev 19(4):327–346

Bellakhal R, Kheder SB, Haffoudhi H (2016c) Institutional and market factors driving renewable energy development in MENA region: a panel data approach. Retrieved October 26th 2018 from Brand B, Zingerle J (2011) The renewable energy targets of the Maghreb countries: impact on electricity supply and conventional power markets. Energy Policy 39(8):4411–4419

Benkhedda M, Boukli-Hacene M, Bouchekara HREH (2020) Renewable energy development in Algeria: current status, barriers, and potential. Renew Energy Focus 35:1–13

Boko M, Niang I, Nyong A, Vogel C, Githeko A, Medany M, Osman-Elasha B, Tabo R, Yanda P (2007) Africa. Climate change 2007: impacts, adaptation and vulnerability. In: Parry ML, Canziani OF, Palutikof JP, van der Linden PJ, Hanson CE (eds) Contribution of working group II to the fourth assessment report of the intergovernmental panel on climate change. Cambridge University Press, Cambridge UK

Carley S (2009) Distributed generation: An empirical analysis of primary motivators. Energy Policy 37(5):1648–1659

da Silva PPM, Cerqueira PA, Ogbe WF (2018) Drivers for renewable energy: a comparison among G20 countries. Energy Rep 4:544–557

Del Río P, Burguillo M (2008a) An empirical analysis of the impact of renewable energy deployment on local sustainability. Renew Sustain Energy Rev 12(9):2498–2509

Del Río P, Burguillo M (2008b) Assessing the impact of renewable energy deployment on local sustainability: towards a theoretical framework. Renew Sustain Energy Rev 12(5):1325–1344

Dragoman D (2014a) Renewable energy policy in Romania: a review. Renew Sustain Energy Rev 36:353–369

Dragoman MC (2014b) Factors influencing local renewable energy initiatives in different contexts: comparative analysis: Italy, Romania and the Netherlands. Master's Thesis, University of Twente

El-Katiri L, Fattouh B (2015)a A roadmap for renewable energy in the Middle East and North Africa. Oxford Institute for energy studies paper MEP 11

El-Katiri L, Fattouh B (2015b) A roadmap for renewable energy in the Middle East and North Africa. Oxford Institute for energy studies paper MEP13. Oxford Institute for Energy Studies, Oxford

El-Katiri L, Fattouh B (2015c) A roadmap for renewable energy in the Middle East and North Africa. Oxford Institute for energy studies paper MEP 11. https://www.oxfordenergy.org/wpcms/wp-content/uploads/2015/01/MEP-11.pdf

Gamel A, Madlener R, Odening M (2017a) The impact of behavioral factors on the adoption of renewable energies by retail investors: an agent-based approach. Energy Policy 109:279–291

Gamel C, Brudermann T, Posch A (2017b) Factors influencing retail investors' attitudes towards investments in renewable energies. J Clean Prod 162:1329–1337

Gamel J, Menrad K, Decker T (2017c) Which factors influence retail investors' attitudes towards investments in renewable energies? Sustain Prod Consump 12:90–103

Hrabanski M, Bidaud C, Le Coq J, Méral P (2013) Environmental NGOs, policy entrepreneurs of market-based instruments for ecosystem services? A comparison of Costa Rica, Madagascar and France. Forest Policy Econ 37:124–132. https://doi.org/10.1016/j.forpol.2013.09.001

IEA (International Energy Agency) (2020) Africa energy outlook 2020: a special report in the world energy outlook series. IEA, Paris

International Energy Agency (2014a) World energy outlook 2014. OECD/IEA, Paris

International Energy Agency (2014b) World energy investment out- look 2014 Special report. Paris

IRENA (International Renewable Energy Agency) (2020) Planning and prospects for renewable power: North Africa. IRENA, Abu Dhabi

IRENA (2021a) Renewable energy statistics. https://www.ren21.net/reports/global-status-report/

IRENA (2021b) Renewable energy outlook: Africa. https://www.ren21.net/reports/global-status-report

Marques AC, Fuinhas JA, Manso JP (2010) Motivations driving renewable energy in European countries: a panel data approach. Energy Policy 38(11):6877–6885

Masini A, Menichetti E (2012) The impact of behavioural factors in the renewable energy investment decision making process: conceptual framework and empirical findings. Energy Policy 40:28–38

Musango JK, Brent AC (2011) A conceptual framework for energy technology sustainability assessment. Energy Sustain Dev 15(1):84–91

Ozturk I, Acaravci A (2013) The long-run and causal analysis of energy, growth, openness and financial development on carbon emissions in Turkey. Energy Econ 36:262–267

Painuly JP (2001) Barriers to renewable energy penetration; a framework for analysis. Renewable Energy 24(1):73–89

Rogers JC, Simmons EA, Convery I, Weatherall A (2012) Social impacts of community renewable energy projects: findings from a woodfuel case study. Energy Policy 42:239–247

Sghaier IM, Madlener R, Odening M (2021) Foreign capital inflows and economic growth in North African countries: the role of human capital. J Knowl Econ. https://doi.org/10.1007/s13132-021-00843-5

Sovacool BK (2009) The importance of comprehensiveness in renewable electricity and energy-efficiency policy. Energy Policy 37(4):1529–1541

Sovacool BK, Drupady IM, Jain A, Abrahamse W (2017) Best practices in energy poverty alleviation: a cross-case comparative analysis. Energy Policy 110:623–634

UNCCD (United Nations Convention to Combat Desertification) (2017) The global land outlook, 1st edn. UNCCD, Bonn

Walker G, Devine-Wright P (2008) Community renewable energy: what should it mean? Energy Policy 36(2):497–500

World Bank (2022) World development report 2022: Finance for an equitable recovery. Washington, DC: World Bank. https://www.afdb.org/fileadmin/uploads/afdb/Documents/Publications/African_Development_Report_2012.pdf. Accessed 23 Jul 2023

Wüstenhagen R, Wolsink M, Bürer MJ (2007) Social acceptance of renewable energy innovation: an introduction to the concept. Energy Policy 35(5):2683–2691

# Chapter 7
# Future Research Directions

**Abstract** This final chapter suggests some policy implications and recommendations that can promote and speed up the deployment of renewable energy sources in Africa. Moreover, it identifies some research gaps that remain to be addressed by future studies on renewable energy and sustainable development in Africa and proposes some possible research questions that future researchers could address to advance our knowledge and practice in this field. The chapter aims to contribute to the understanding of the dynamics of renewable energy in Africa, as well as to the policy design and implementation for renewable energy development and integration in the continent.

*What is in for the readers of this chapter*? The readers can explore some of the policy implications and recommendations that can promote and speed up the deployment of RE sources in Africa.

## 7.1  Introduction

Renewable energy is widely recognized as a key driver of sustainable development in Africa, as it can provide clean, affordable, and reliable electricity to millions of people who currently lack access to modern energy services. Renewable energy can also reduce greenhouse gas emissions, enhance energy security, diversify the energy mix, create jobs, and foster innovation. However, renewable energy development and utilization in Africa face many challenges, such as technical, financial, institutional, regulatory, and social barriers, as well as knowledge and data gaps. Therefore, there is a need for more research on the impacts of renewable energy on sustainable development in Africa, and how to overcome the existing challenges and seize the opportunities.

This book aims to fill some of these knowledge and data gaps by providing an overview of the electrification challenges and opportunities in Africa, with a focus on renewable energy sources, such as geothermal, solar, wind, hydro, and biomass. The

© The Author(s), under exclusive license to Springer Nature Switzerland AG 2024
H. Qudrat-Ullah, *Exploring the Dynamics of Renewable Energy and Sustainable Development in Africa*, Advances in African Economic, Social and Political Development, https://doi.org/10.1007/978-3-031-48528-2_7

book also presents case studies of four African countries, namely Cameroon, Nigeria, Uganda, South Africa, and Algeria, that have different energy profiles, potentials, and contexts. The book analyzes the electricity demand and supply, the policy and technical solutions, the socioeconomic benefits and environmental impacts, and the barriers and gaps in renewable energy development and utilization in these countries. The book also suggests some policy implications and recommendations that can facilitate and accelerate the deployment of renewable energy sources in Africa.

In this final chapter of the book, we identify some research gaps that remain to be addressed by future studies on renewable energy and sustainable development in Africa. We also propose some possible research questions that future researchers could address to advance our knowledge and practice in this field. We hope that this book will stimulate further research on this topic and inspire more action toward achieving universal access to clean and affordable electricity in Africa.

## 7.2  Future Research Direction for RE and Sustainable Development in Africa

In this book, we have explored the electrification challenges and opportunities in Africa from a renewable energy perspective. We have also examined the case studies of four African countries with different energy profiles, potentials, and contexts: Cameroon, Nigeria, Uganda, South Africa, and Algeria. We have assessed the electricity demand and supply, the policy and technical solutions, the socioeconomic benefits and environmental impacts, and the barriers and gaps of renewable energy development and utilization in these countries. We have also proposed some policy implications and recommendations that can promote and speed up the deployment of renewable energy sources in Africa. These include creating an enabling business environment, enhancing local technical capability and research and development, improving data and statistics, diversifying the energy mix and reducing dependence on fossil fuels, and enhancing electricity access, efficiency, and sustainability. However, as is the case of any research project, there are still many research opportunities for future studies. Some of these research gaps include:

1. The lack of comprehensive and integrated assessment frameworks and indicators that can capture the multiple and interrelated impacts of renewable energy on sustainable development across different scales, sectors, and contexts in Africa. Some interesting research questions that future researchers could address include:

   - What are the most effective ways to integrate renewable energy into existing infrastructure systems such as transportation water supply waste management etc.?
   - How can we develop comprehensive and integrated assessment frameworks and indicators that can capture the multiple and interrelated impacts of renewable energy on sustainable development across different scales, sectors, and contexts in Africa?

- What are the most effective innovative financing mechanisms for renewable energy projects such as green bonds crowdfunding microfinance etc.?
- How can we collect empirical evidence and data on the long-term and cumulative impacts of renewable energy on sustainable development in Africa especially on the environmental and social dimensions as well as the feedback loops and interactions between different dimensions?

2. The lack of empirical evidence and data on the long-term and cumulative impacts of renewable energy on sustainable development in Africa, especially on the environmental and social dimensions, as well as the feedback loops and interactions between different dimensions. Some possible research questions that future research are:

- How do different renewable energy sources (such as solar, wind, hydro, etc.) affect the environmental and social dimensions of sustainable development in Africa? What are the trade-offs and synergies between them?
- How can renewable energy policies and regulations in Africa be designed to promote social inclusion, gender equality, poverty reduction, and human rights?
- How can renewable energy projects in Africa enhance local capacities, create jobs, foster innovation, and support community empowerment?
- How can renewable energy integration in Africa's power systems improve energy access, reliability, affordability, and resilience?
- How can regional cooperation and integration in renewable energy development and trade in Africa foster peace, security, and economic growth?

3. The lack of comparative analysis of the impacts of different types of renewable energy technologies, such as geothermal, solar, wind, hydro, biomass, etc., on sustainable development in Africa, taking into account their technical, economic, social, and institutional characteristics and implications. The following research questions are up for future research.

- What are the technical, economic, social, and institutional characteristics and implications of different types of renewable energy technologies such as geothermal, solar, wind, hydro, biomass, etc. on sustainable development in Africa?
- What are the comparative impacts of different types of renewable energy technologies such as geothermal, solar, wind, hydro, biomass, etc. on sustainable development in Africa?
- How can we develop comprehensive and integrated assessment frameworks and indicators that can capture the comparative impacts of different types of renewable energy technologies on sustainable development across different scales sectors and contexts in Africa?

4. The lack of participatory and inclusive research methods that involve the perspectives and experiences of various stakeholders, especially the local communities and groups that are affected by or benefit from renewable energy projects in

Africa. Some emerging research questions that future researchers could address include:

- How can we develop participatory and inclusive research methods that involve the perspectives and experiences of various stakeholders, especially the local communities and groups that are affected by or benefit from renewable energy projects in Africa?
- What are the best practices for involving local communities and groups in renewable energy projects in Africa?
- How can we ensure that renewable energy projects in Africa are socially inclusive and equitable?

5. The lack of policy analysis and evaluation studies that examine the effectiveness and efficiency of existing policies and programs that support or hinder renewable energy transition in Africa about sustainable development goals and targets. Some RE advancing in Africa research questions include:

- What are the existing policies and programs that support or hinder renewable energy transition in Africa about sustainable development goals and targets?
- What are the best practices for policy analysis and evaluation studies that examine the effectiveness and efficiency of existing policies and programs that support or hinder renewable energy transition in Africa regarding sustainable development goals and targets?
- How can we ensure that existing policies and programs that support or hinder renewable energy transition in Africa are socially inclusive and equitable?

These research gaps indicate the need for more rigorous and comprehensive studies on the impacts of renewable energy on sustainable development in Africa that can inform and support policymaking and practice in this field. Such studies can also contribute to the global knowledge base on renewable energy and sustainable development and provide valuable insights and lessons for other regions and countries that are undergoing or planning to undergo similar transitions. The book hopes that its findings and recommendations can stimulate further research on this topic and inspire more action toward achieving universal access to clean and affordable electricity in Africa.

In conclusion, this book has identified some of the key challenges and opportunities of renewable energy development and utilization in Africa. The book has also suggested some policy implications and recommendations that can facilitate and accelerate the deployment of renewable energy sources in Africa. Future research could address the research gaps identified in this book to further advance our understanding of renewable energy's impact on sustainable development in Africa.

## 7.3 My Words About This Book

I hope that this book has offered a useful overview of how renewable energy sources such as geothermal solar wind hydro biomass can address the electrification challenges and opportunities in Africa. I also hope that the case studies of Cameroon, Nigeria, Uganda, South Africa, and Algeria have shown the diversity and complexity of renewable energy development utilization in different contexts and settings. I acknowledge that there are still many research gaps and challenges that need to be addressed by future studies on renewable energy and sustainable development in Africa. I encourage future researchers to pursue the research questions and directions suggested in this book, as well as to explore new and innovative ways of researching this topic. I believe that renewable energy can play a vital role in achieving sustainable development in Africa, as well as in the rest of the world. I also believe that renewable energy can bring multiple benefits to millions of people who currently lack access to modern energy services, such as improved health, education, livelihoods, and well-being. I conclude by expressing my gratitude to all the contributors, reviewers, editors, and readers of this book for their valuable input, feedback, support, and interest. I humbly hope that this book will inspire more action and collaboration towards achieving universal access to clean and affordable electricity in Africa and beyond.

# Index